Money, Greed, and God

Money, Greed, and God

Why Capitalism Is the Solution and Not the Problem

Jay W. Richards

HarperOne
An Imprint of HarperCollinsPublishers

HarperOne

HarperCollins Web site: http://www.harpercollins.com
HarperCollins®, ®, and HarperOne™ are
trademarks of HarperCollins Publishers

Designed by Level C

FIRST EDITION

Richards, Jay Wesley.
  Money, greed, and God : why capitalism is the solution and not the problem / by
Jay Richards.—1st ed.
    p. cm.
Includes bibliographical references.
ISBN 978–0–06–137561–3
1. Capitalism—Religious aspects—Christianity. I. Title
BR115.C3R53 2009
261.8'5—dc22

                              2008051785

    09 10 11 12 13 RRD (H) 10 9 8 7 6 5 4 3 2 1

To Gillian and Ellie

Contents

Figures and Tables

Money, Greed, and God

Can a Christian Be a Capitalist?

In the great twentieth-century battle between communism and capitalism, capitalism won. But many of us still have serious problems with capitalism. Just turn on the TV and you'll see capitalism blamed for almost every social problem. Executives at giant corporations like Enron, WorldCom, and Tyco lie, cheat, and steal billions of dollars from workers and shareholders. A few lucky people get fabulously wealthy while grinding poverty persists, not just in third-world countries, but in the middle of great American cities. Oil companies despoil the environment, use up limited natural resources, and resist efforts to find alternative sources of energy. As one anonymous critic put it: "The history of capitalism is a history of slavery, child labor, war, and environmental pollution."

But even if we write all of this off as media bias and bad philosophy, there is still our everyday experience: Pressure from bosses and stockholders to increase profits at any cost. The frantic race to keep up with the Joneses that never seems to stop. Factories that close and leave entire towns high and dry. Ugly billboards that spoil cityscapes and landscapes. Advertising companies that find ever more sophisticated ways of luring us to the mall and the Internet to buy kitschy plastic stuff we don't need, that nobody needs. Landfills piled high with computer monitors and keyboards that still work. You don't have to be a snooty leftist to find all of this distasteful.

Add to the bad press and our everyday experience the views of some *champions* of capitalism, who defend the system wholly on the grounds of self-interest. Over two hundred years ago, the founder of modern economics, Adam Smith, said, "It is not from the benevolence of the butcher, the brewer, or the baker that we expect our dinner, but from their regard to their own interest." That's not a repugnant comment, but it's not morally uplifting, either. More recently, the atheist Ayn Rand, revered by millions as the greatest defender of capitalism, seemed to take Smith's reasoning to the extreme. She claimed that greed—which she called a virtue—is the basis for a free economy. Christian altruism, in contrast, is capitalism's bitter enemy.

Rand's philosophy reached a high-water mark in 1986, when investor Ivan Boesky told graduates at the University of California at Berkeley in a commencement speech: "Greed is all right, by the way . . . I think greed is healthy. You can be greedy and still feel good about yourself." If Rand and her disciples like Boesky are right, this creates a big problem for Christians in capitalist societies.

Yet, as a practical matter, we take capitalism for granted. We live and breathe it every day. We own cars and furniture. We work for and own businesses. We try to get the best salary we can. We take out mortgages to finance our houses and pay interest on the loan every month. We use Visa, MasterCard, and American Express. On eBay, we sell to the highest bidder, and we seek the best deal when we're bidding on a home gym or that latest Hummel figurine we simply must have. We may even own Exxon-Mobil stock and Vanguard mutual funds. It's hard to imagine an alternative. The few remaining communists are cloistered mostly in places like Harvard and Havana.

But history's harsh verdict on communism has not led to a warm Christian embrace of capitalism. On the contrary, we're told that capitalism destroys our natural and moral environment. We hear this on TV and in newsprint, of course; but we also hear it in churches and Bible studies. Churchgoing businesspeople endure sermons that show little understanding of business. Many thoughtful Christians suffer pangs of doubt. Or they feel vaguely guilty, even when their experience contradicts what they hear from the pulpit. We hear about "business ethics," as if business had unique moral hazards not encountered in the law or medicine or education or the ministry. Few have ever heard a sermon that dignifies business as a calling. Few have heard their pastor or priest extol the virtues of the free market for spreading prosperity. Most assume that they just have to live with the tension.

Of course, there is the opportunistic "name it and claim it" prosperity gospel, whose evangelists preach that God wants everyone to be rich, RICH, RICH. (I guess the twelve apostles didn't get the memo). But this view remains the exception rather than the rule; and it's an exception that makes it all the easier for most Christian intellectuals to hold capitalism at arm's length or even reject it. The popular book by evangelical Jim Wallis, *God's Politics: Why the Right Gets It Wrong and the Left Doesn't Get It,* is one recent example. While Wallis admits that the market economy is efficient and perhaps unavoidable, he nevertheless attacks capitalism on moral and biblical grounds. More and more Christians share his views.

I used to be one of them. In fact, for a while I hated capitalism. As a young college student in the 1980s, I struggled with the apparent conflict between Christ's ample warnings about money and the rank materialism of American life. When I read the Bible, it seemed to brim with passages designed to make the capitalist squirm:

"You cannot serve both God and Mammon," Jesus said. *Well, then,* I thought, *where does that leave the almighty dollar?*

"It is easier for a camel to pass through the eye of a needle than for a rich man to enter the kingdom of heaven." *Well, then, judging by historical standards, most of the American middle and upper classes probably won't get to heaven.*

"Do not lay up for yourselves treasures on earth, where moth and rust destroy and where thieves break in and steal; but lay up for yourselves treasures in heaven, where neither moth nor rust destroys and where thieves do not break in and steal. For where your treasure is, there your heart will be also." *That seems to scotch all those well-laid plans to invest and save for the future.*

"The love of money is the root of all evil." *But isn't capitalism based on the love of money?*

These interpretations seemed right at the time. As a college freshman in 1985—before the full results were in from the communist experiment in the Soviet Union—I had a vaguely Christian and clearly sophomoric infatuation with Marxism. When I learned about the many problems with Marxism, I moved on to democratic socialism. That sounded much nicer. The more serious I became as a Christian and as a student of economics, however, the more I came to believe that capitalism was practically if not morally superior. This is where many thoughtful Christians end up. But after several years of graduate school and still more struggle, I eventually discovered that capitalism could be soundly defended on both practical and moral grounds.

To be sure, if every bad thing we're led to believe about capitalism were true, I would join the critics at the front of the line. But the problems we hear about, the ones that shape how we see capitalism, are not the whole story. They're not even the real story. Capitalism will not usher in a utopia, of course. We're all sinners, even in the best of circumstances. Get a bunch of us together, and there's bound to be plenty of sin all around. But many Christians confuse a caricature of capitalism with the real thing. To make matters worse, we often make serious mistakes when we turn our attention to economics.

If we want to know the truth, though, if we want to order our economic and social lives justly, if we want to help people rather than merely feel like we're helping people, we have to learn the economic terrain. Rich Karlgaard, the Christian publisher of *Forbes* magazine, has complained that listening to a pastor or priest preach on business "is like hearing a eunuch lecture on sex: He may have studied the topic but really knows little about the mechanics."[1] Why is that? No serious Christian writing about natural science would ignore the basic facts of chemistry or astronomy. But too many Christian leaders feel free to ignore the basic facts of economics. This is a serious mistake, because economics is not just a bundle of baseless opinions. The twentieth century was one big lab experiment for the economic theories of the eighteenth and nineteenth centuries. The results of that experiment are in. It's time for Christians to take an honest look at the economic facts and get acquainted with capitalism.

The first order of business is to clear out the fuzz and fog in our thinking. This is a hurdle, since it means we have to do more than worry or care deeply. We have to *think*. Fortunately, we don't need to study supply-and-demand charts and complicated statistics to do that. At bottom, economics is about us—what we choose, what we value, what we represent in language and symbols, how we interact with each other in a market, and especially how we produce, exchange, and distribute goods, services, risk, and wealth. Understand these things, and you're well on your way to knowing what you need to know about economics.

If you think about these matters at all, you probably have asked questions like the following:

Can't we build a just society?

What does God require of us as Christians?

Doesn't capitalism foster unfair competition?

If I become rich, won't someone else become poor?

Isn't capitalism based on greed?

Has Christianity ever really embraced capitalism?

Doesn't capitalism lead to an ugly consumerist culture?

Do we take more than our fair share? That is, isn't our modern lifestyle causing us to use up all the natural resources?

These are all good questions. To answer them correctly, however, we have to separate them from common economic mistakes that easily creep into our thinking. That's our second order of business.

Fortunately, almost every mistake that Christians make in economics can be boiled down to eight simple myths. I'm intimately familiar with them, since they distorted my own thinking at one time:

1. The Nirvana Myth (contrasting capitalism with an unrealizable ideal rather than with its live alternatives)

2. The Piety Myth (focusing on our good intentions rather than on the unintended consequences of our actions)

3. The Zero-Sum Game Myth (believing that trade requires a winner and a loser)

4. The Materialist Myth (believing that wealth isn't created, it's simply transferred)

5. The Greed Myth (believing that the essence of capitalism is greed)

6. The Usury Myth (believing that working with money is inherently immoral or that charging interest on money is always exploitive)

7. The Artsy Myth (confusing aesthetic judgments with economic arguments)

8. The Freeze-Frame Myth (believing that things always stay the same—for example, assuming that population trends will continue indefinitely, or treating a current "natural resource" as if it will always be needed)

Each of these myths is easy to expose in the abstract, but it's equally easy to forget, in practice, that they're myths. Remember that they are, however, and you'll be immunized against a lot of economic folderol, even if you never take a course in economics.

Finally, we have to think about capitalism as Christians. The great Dutch statesman and theologian Abraham Kuyper once said that "there is not a square inch in the whole domain of our human existence over which Christ, who is Sovereign over all, does not declare, 'Mine!'" This holds for economics as well. Thinking as a Christian doesn't mean we just slap some Bible verses on our economic views so that they sound biblical. It means taking the core truths of the faith and using them as a lens to cast new light on unexplored territory.

Economic truths are truths. But they don't stand outside God's dominion. Being a Christian doesn't mean you can disregard economic facts. But understanding economics doesn't answer the really tough questions: Does Christian theology have anything to contribute to capitalism? Does capitalism fit with the Christian view of the world? And the big one: Is the capitalist system consistent with Christian morality? If a free-market economy contradicts Christian ethics, Christians can't be capitalists.

As it turns out, there is no such contradiction. We suspect one only because many of capitalism's champions *and* critics miss the subtleties of the capitalist system: to prosper, a market economy needs not just competition, but rule of law and virtues like cooperation, stable families, self-sacrifice, a commitment to delayed gratification, and a willingness to risk based on a future hope. These all fit nicely with the Christian worldview.

Moreover, despite what you've been told, the essence of capitalism is not greed. It's not even competition, private property, or the pursuit of rational self-interest. These last three items, rightly defined, are key ingredients in any market economy; but the heart of capitalism lies elsewhere. What we now know is that market economies work because they allow wealth to be *created*, rather than remaining a fixed pie. Economies need not be zero-sum games in which someone wins only if someone else loses. We have discovered an economic order that creates wealth in abundance—capitalism. And only the creation of wealth will reduce poverty in the long run.

How is wealth created? The economist can't easily answer this question, but the Christian *can*. Paradoxically, the key source of material wealth in a modern market economy is immaterial. It's spiritual. Wealth is created when our creative freedom is allowed to prosper in a free-market environment undergirded by the rule of law and suffused with a rich moral culture. This creative freedom should be no surprise to Christians. We believe that human beings are made in God's image—the *imago dei*. Our creative freedom reflects that divine image. This is one of the least appreciated truths of economics.

When we have all the facts before us, then, we discover that Christianity and capitalism are not bitter enemies. On the contrary, a good Christian can be, indeed should be, a good capitalist. If you're not yet convinced, keep reading.

One

Can't We Build a Just Society?

BIG CARS, BIG STEAKS, BIG CHURCHES

As a young adult, I liked revolutionary rhetoric. It was more sophomoric rebellion than careful conclusion, but it was enough. In my eighteen-year-old mind, I had plenty to rebel against. I was born and raised in Amarillo, a place on the High Plains of the Texas Panhandle so isolated we thought our town of 150,000 was the big city. Three icons capture the flavor of Amarillo better than an entire essay. It is bounded on the west by the Cadillac Ranch—ten Cadillacs planted nose down in concrete, their tail fins neatly in a row—and on the east by the Big Texan Steakhouse, a restaurant where you can get a seventy-two-ounce steak free if you can eat it in one hour. In between are a bunch of really big churches. In four blocks of downtown Amarillo, for instance, the First Baptist Church, the Central Church of Christ—where my best friend went—and St. Mary's Catholic Church together claimed more than twelve thousand members.

In between these parking-lot-happy congregations sat our church, First Presbyterian. It was smaller than the others, but with its quasi-Gothic architecture, gray stones, and red tile roof, it looked like a country-club church—a problem exacerbated by the fact that it *was* a country-club church. Our congregation heard few hard lies from the pulpit—and even fewer hard truths.

It was a comfortable church, a church that found the culture wars unseemly.

But things changed when I entered junior high and came under the sway of a firebrand youth pastor, Darren Clark. Clark was a divorced former Southern Baptist (a core Presbyterian constituency in Texas) who had an uncanny knack for illustrating Bible stories with lyrics from the Beatles and Simon and Garfunkel. I wasn't always sure if his lessons owed more to the Beatles or the Bible, but they were easy to remember.

He was always railing against nuclear weapons, war, and the indifference of Christians to poverty. Whether Jesus died for our sins was less of a priority. In any case, he must have railed a lot against the evils of wealth, because I remember the Bible verses on those topics better than any others: "You cannot serve both God and Mammon." "It is easier for a camel to pass through the eye of a needle than for a rich man to enter the kingdom of heaven." "Do not lay up for yourselves treasures on earth, where moth and rust destroy and where thieves break in and steal; but lay up for yourselves treasures in heaven, where neither moth nor rust destroys and where thieves do not break in and steal. For where your treasure is, there your heart will be also." "The love of money is the root of all evil."

New thoughts entered my mind: Is it right that some should be so rich while others remain poor? Should hordes of people have to work in unsafe conditions, barely getting by on their meager salaries, while their bosses grow fat and complacent? Wouldn't it be better if everyone had as much as they want, or at least as much as they need? Shouldn't we do whatever is necessary to get rid of such injustices? These questions bothered me. I wasn't sure what we were supposed to do about poverty or nukes other than feel really bad about them, but still, to a teenager in a conservative Texas town, this was heady stuff. Clark offered me a rare commodity: a chance to rebel against authority and feel self-righteous doing it.

EXIT STAGE LEFT

The seeds sown in junior high and high school were reaped in college. I attended a small liberal-arts school north of Austin, Southwestern University. The college had a tenuous tie to the Methodist Church, but whatever Christianity remained had long since been converted to Protestant Liberalism—think of the *New York Times* editorial page laced with Bible verses. To a freshman weaned on George Strait and chicken-fried steak, the ambiance seemed smart and sophisticated.

I had already decided to study political science, so I found myself right away in "Introduction to American Politics," taught by a soft-spoken professor, Suk Kim, who had immigrated to the United States from Korea. In one of his first lectures, Kim asked the class to name the most disagreeable jobs we could think of. Someone said garbage collector, another said sewer repair, then hard laborer in extreme climates, and so forth. He then asked for the most agreeable jobs, which, everyone agreed, were movie stars, professional athletes and musicians, CEOs of large corporations, bank presidents, and so forth. Kim then asked us to rank the pay scales of the various professions. With few exceptions, the jobs voted most disagreeable were also the lowest paying. Kim asked to great effect if anyone thought that was fair. No one raised a hand. I certainly didn't. The lesson was obvious.

Professor Kim's lectures were reinforced by the reading assignments. He required a standard textbook describing the bicameral legislature, the separation of powers into the judicial, legislative, and executive branches, and all that other stuff I had already learned in high school civics. But we had two other, quite memorable books. The first was a short work by Karl Marx and Friedrich Engels—*The Communist Manifesto*. Written in 1848, it describes history as a series of struggles between oppressors and oppressed, each struggle punctuated by a social upheaval in which one system gives rise to another. According to the *Manifesto,* the original state of man was a primitive communism without private property. The mass slavery of Egypt and other ancient cultures represented the fall from this primeval state of innocence. The slave system eventually gave way to feudalism, with poor serfs caring for large tracts of land owned by nobles. And feudalism eventually gave way to capitalism, with its highly productive urban industrial centers.

In modern capitalist societies, Marx and Engels argued, the business owners—what they call the bourgeoisie—seek above all else to increase their profits. So they pay workers as little as possible while taking the "surplus value" produced by the workers as profit. With those profits, they then invest in tools and factories to extract more from the labor of fewer laborers, which again creates more surplus value. The capitalists compete to produce more and more with less and less. Workers become more and more alienated from the fruits of their labor, as capitalists skim more and more of the surplus value of their work without giving them just compensation.

The bigger and better capitalists inevitably beat out the smaller ones. These big businesses can then produce far more with still fewer workers. In this way the few remaining business owners inevitably grow fatter and richer, while a growing pool of laborers grows poorer and poorer. Capitalism thus sows the seeds of its own destruction, by creating a large oppressed population that will eventually revolt against its oppressors. The workers, Marx and Engels predicted, would usher in the "dictatorship of the proletariat," where the people—that is, the state—would own all industry. But for those afraid of big government, don't worry. This would be merely a temporary socialist way station on the road to full communism, where the state would wither away into a brotherhood of man, and milk and honey would flow through streets of gleaming gold, or something like that.

Although Marx and Engels claimed that such a revolution is inevitable, they wrote *The Communist Manifesto* to help the inevitable along. They concluded with a call to arms:

> The Communists disdain to conceal their views and aims. They openly declare that their ends can be attained only by the forcible overthrow of all existing social conditions. Let the ruling classes tremble at a Communist revolution. The

proletarians have nothing to lose but their chains. They
have a world to win.

Workingmen of all countries, unite!

This worked well alongside the other required text, *Democracy for the Few,* by American socialist Michael Parenti. With
Parenti's book, Professor Kim offered students a great deal: read
it five times and get an "A" for the course.[1] For me, it was an easy
assignment, since the book was filled with facts I couldn't refute
and moral fury I couldn't resist. On almost any page, Parenti offered hard evidence of the perverse inequalities of American life:

> The top 10 percent of American households own 98 percent
> of the tax-exempt state and local bonds, 94 percent of business assets, and 95 percent of the value of all trusts. The
> richest 1 percent own 60 percent of all corporate stock and
> all business assets. True, some 40 percent of families own
> some stocks or bonds, but almost all of these have total
> holdings of less than $2,000. Taking into account their
> debts and mortgages, 90 percent of American families have
> little or no net assets.

By the end of the semester, I had highlighted practically every
line in the book. Parenti's point seemed undeniable: Just as Marx
and Engels had predicted 150 years earlier, American capitalism had produced gross inequalities between rich and poor. A
rich oligarchy controlled more and more of the wealth and left
the vast majority of Americans with very little. The only way to
balance the scales was for the national wealth to be owned and
controlled by the people as a whole.

My five trips through Parenti's book gained me zeal and an
"A" for the course. I had a grid for interpreting pretty much
everything. Suddenly, the materialism and greed of American
life started popping up everywhere. I noticed all the BMWs and
Saabs on campus. And I was surprised how many guys in my
dorm had the same poster of Garfield the cat. On the poster,
Garfield was covered in jewels and leaning against a Ferrari,

with a little proverb at the bottom: "He who dies with the most toys, wins." That pretty much summed it up. That was capitalism. I could imagine what Jesus would do if he ever showed up. He'd bring the whip he used against the money changers and put it to good use. Nobody cared about the poor more than he did. And the Bible speaks over and over about a time when the rich and haughty would be brought low, and the poor lifted up:

> Those who were full have hired themselves out for bread,
> but those who were hungry are fat with spoil. . . .
> He raises up the poor from the dust;
> and lifts the needy from the ash heap. (1 Sam. 2:5, 8)

Marx's atheism was still problematic for me, but I figured it wasn't essential to Marxism. I mean, the rest of the communist picture seemed to fit nicely with the early church in Acts, where "the believers were together and had everything in common. Selling their possessions and goods, they gave to anyone as he had need" (Acts 2:44–45). I concluded that Marx had gotten it right in spite of his atheism, not because of it.

But a funny thing happened on the way to the Marxist utopia—a messy little thing called reality. That reality is different in ever particular from the practice of the early church in Jerusalem.

THE SPACE BETWEEN REALITY AND RHETORIC

It was 1985, and the Cold War between the Western democracies and the communist Soviet Union was in its final stages, though no one knew it at the time. The news media were wringing their hands about the nuclear arms race, made-for-TV movies like *The Day After* were forecasting the dismal fate of the earth in the wake of a nuclear holocaust, and just about everyone except Ronald Reagan and Margaret Thatcher thought the Soviet Union was here to stay.

Oops: "Can economic command significantly . . . accelerate the growth process? The remarkable performance of the Soviet Union suggests that it can. . . . Today the Soviet Union is a country whose economic achievements bear comparison with those of the United States."

—MIT economist
Lester Thurow, 1989

A humane leader had emerged in the Soviet Union, a man named Mikhail Gorbachev. He was implementing policies that would soon lead to the collapse of the Soviet empire. While his predecessors had preferred varying degrees of secrecy both with Soviet citizens and with the outside world, Gorbachev promoted "glasnost" and "perestroika." Suddenly, Soviet history since the 1917 communist revolution was open for all to see. It wasn't pretty.

Of course, communist crimes weren't exactly a well-kept secret before this. At one time in the mid–twentieth century, almost half the human race was subject to a grand Marxist experiment. So half of humanity knew the truth firsthand, even if much of the other half chose to ignore it. Already by the 1970s, the results were in for anyone willing to do a little homework. But it was not until the Gorbachev era that the Soviet regime began to admit to the world what many had known all along: whatever idealistic vision may have inspired it, the commune in communism was one nasty neighborhood. This made it much harder for apologists in the West to keep up the charade.

Marx had predicted that the contradictions in capitalism would eventually cause the workers to revolt. He was quite clear that the growth of wealth and industrial productivity in capitalism was a crucial stage on the way to future stages of social evolution. Only after capitalism's tensions grew to the breaking point would the workers revolt and usher in a socialist state, where private property would be abolished. Even in Marx's lifetime, however, his prophecies clashed with reality. He spent his later years in England writing, but never completing, his master-work, *Capital*. And while he scribbled away in his study, labor-ers' wages in England were rising, not falling. Marx apparently didn't notice.[2]

When a communist revolution finally succeeded in 1917, it was led by intellectuals in an agrarian culture that had little his-tory with either democracy or capitalism. As Harvard historian Richard Pipes puts it: "Communism . . . did not come to Russia as the result of a popular uprising: it was imposed on her from above by a small minority hiding behind democratic slogans."[3] Contrary to Marx's predictions, this was the pattern of commu-nist revolutions throughout the twentieth century.

The 1917 Russian Revolution was led by an angry and fa-natical intellectual named Vladimir Lenin. He led his Bolshevik ("majority") Party to victory after a three-year civil war. Before the revolution, Russia had been ruled by a czar. Russian society was divided between a small elite aristocracy and a large popula-tion of rural peasants, with few capitalists to speak of.

Lenin quickly abolished all legal hindrances to his absolute rule and set up a one-party system in which the Bolshevik Party (soon renamed the Communist Party) filled every nook and cranny of Russian society. He began to centralize large chunks of the Russian economy, from industry and trade to education and transportation. This required secret police, a massive bu-reaucracy, and the widespread use of terror.

Lenin's attempts to centralize the economy were utter di-sasters. The dictatorship of the proletariat quickly became the dictatorship of recalcitrant bureaucrats. To his chagrin, Lenin found that bureaucrats in Moscow were neither motivated nor competent to manage distant factories and farms. Restrictions on trade created a black market that was larger than the official economy. To add insult to injury, the regime dumped banknotes into the market, which predictably led to runaway inflation. By 1923, prices were 1 million times greater than prices before the revolution began.

Across the economy, productivity plummeted: "Overall large-scale industrial production in 1920 was 18 percent of what it had been in 1913. . . . The number of employed industrial workers in 1921 was less than one-half of what it had been in 1918; their living standard fell to one-third of its prewar level."[4]

Agriculture was even worse. Lenin tried to force peasants to sell their grain below market price even as he ordered a large-scale massacre of the wealthier peasants, the kulaks. This led to food shortages and massive strikes, which Lenin punished with poison gas. The situation became so dire that in 1921 Lenin instituted the New Economic Policy (NEP), which allowed the peasants to sell their grain for market prices after paying a tax. He also eased some of the restrictions on trade while continuing to assert control of other parts of the economy. These modest reforms allowed grain production to rebound. But it was too late to prevent a famine, brought on by drought, that killed 5.2 million people.

"Marxism has not only failed to promote human freedom, it has failed to produce food."
—American novelist
John Dos Passos

Lenin did not live to see his policies through. That was left to his successor, Joseph Stalin. Under Stalin, Communist Russia quickly absorbed the countries on its border such as Ukraine, and in 1924 Russia formed the Union of Soviet Socialist Republics. Stalin implemented a series of "Five Year Plans" to take control of large sectors of the economy. The livelihoods of industrial workers were decimated, while millions of peasants died from a forced famine in 1932 and 1933. Combined with various purges of Communist Party officials, Stalin orchestrated the largest-scale massacre of a domestic population in human history. At its height in 1937 and 1938, there were *on average* one thousand political executions per day, not including the countless millions sent to labor camps.[5]

Such tragedy was not the exception but the rule for other communist experiments in the twentieth century. Whatever Marx expected, revolutions never sprang up in advanced industrial societies where there was a strong rule of law, but rather in poor agrarian cultures with career tracks for despots.

> "The vicissitudes of history . . . have not dissuaded [leftist intellectuals] from their earnest search for a 'third way' between socialism and capitalism, namely, socialism."
> —Richard John Neuhaus

The Chinese Revolution led by Mao Tse-tung in 1948 differs from the Russian Revolution in details, but the basic plot line is the same: labor and reeducation camps, mass killings, and economic ruin following attempts to collectivize industry and agriculture. Where Stalin had his Five Year Plans, Mao had the "Great Leap Forward": "We shall teach the sun and moon to change places," read one piece of promotional literature. "We shall create a new heaven and earth for man." Perhaps it sounds nice in Mandarin, but more than 20 million Chinese died in the famine that resulted from the heaven-on-earth construction project. Another 20 million died in *laogai,* the Chinese labor camps.[6]

China and Russia take first and second place when it comes to total deaths. But for the prize of applying the brutal logic of equality, no one beats the French-educated Pol Pot and his Khmer Rouge, which ruled Cambodia from 1975 to 1979. No other regime ever worked so hard to create an egalitarian society.

By the time I was a sophomore in college, I knew about the brutalities in Russia and China, and I had even read Aleksandr Solzhenitsyn's grim firsthand record of Soviet labor camps, *The Gulag Archipelago.* But I had never heard of the Khmer Rouge, even though I was studying political science. (What little I had heard of Cambodia came from American linguist Noam Chomsky, who for years denied what happened in the country, and then, when it became undeniable, blamed the United States.)[7]

In 1986, however, I saw *The Killing Fields,* a film based on the story of *New York Times* reporter Sydney Schanberg (played by Sam Waterston) and his Cambodian interpreter, Dith Pran. The movie begins in 1973, with Schanberg and Pran investigating an accidental bombing of a Cambodian village by an American B-52. But most of the film takes place two years later, when the Khmer Rouge overrun the Cambodian capital of Phnom Penh.

Because the Cambodians are at risk of arrest, Schanberg gets evacuation orders for Pran and Pran's wife and children; but after his family is evacuated, Pran decides to stay behind to help. He ends up getting arrested. Then the horror really begins.

To erase the inequalities of the previous order, Pol Pot's regime declares a "Year Zero," renames the country Democratic Kampuchea, and abolishes money. Two and half million occupants of the capital, including women, children, and hospital patients, are ordered to abandon everything and march into the countryside. In one week, Cambodia's major cities are empty: 4 million people are sent to work in labor camps. One of them is Dith Pran.

Pran languishes for months on a communal farm, teetering on the edge of starvation. The new government quickly begins executing everyone who might resist its plans, targeting especially the skilled and well educated. The Khmer Rouge has already trained executioners, picking those least likely to be polluted by their earlier way of life: children. Dith Pran narrowly escapes such an execution-by-child and eventually manages to escape.

Shortly after his escape, he stumbles into a wet rice paddy filled with thousands of human skeletons. He has found one of the infamous "killing fields" of Cambodia, mass graves of the victims of the Khmer Rouge.

Several times the film switches from Pran to Schanberg, now safely back in New York, who tries to blame the horrors of the Khmer Rouge on an overreaction to American bombings. But such leftist glosses, believable only in New York, can't justify the evil of the Khmer Rouge. In forty-four short months, the regime pruned the Cambodian population by about 2 million—more than a quarter of the total population.[8]

Dith Pran was one of the lucky few. He eventually escaped across the border into Thailand and joined his family in the United States.

In the film, after Pran's escape, Schanberg comes to meet him at a Red Cross station near the Thai border with Cambodia. As they embrace, John Lennon's "Imagine" is playing on a radio in the background:

> Imagine there's no Heaven
> It's easy if you try
> No hell below us
> Above us only sky
> Imagine all the people
> Living for today
>
> Imagine there's no countries
> It isn't hard to do
> Nothing to kill or die for
> And no religion too
> Imagine all the people
> Living life in peace
>
> You may say that I'm a dreamer
> But I'm not the only one
> I hope someday you'll join us
> And the world will be as one
>
> Imagine no possessions
> I wonder if you can
> No need for greed or hunger
> A brotherhood of man
> Imagine all the people
> Sharing all the world
>
> You may say that I'm a dreamer
> But I'm not the only one
> I hope someday you'll join us
> And the world will live as one.

The scene was meant to be moving, and it was. But it was also dissonant. Lennon's anthem describes a sort of godless kingdom of God. We're asked to imagine a future "brotherhood of man" where "all the people" live "life in peace." Who wouldn't want that? But read the fine print. This is a world with "no possessions," where "all the people" share "all the world." Huh? Is it just me, or does that sound a lot like the dream of those antireligious communist visionaries that Dith Pran only narrowly escaped? Don't Lennon's words express the same sentimental delusions that inspired communism in the first place? That thought haunted me, though it was years before I was ready to consider a real alternative.

"Socialism in general has a record of failure so blatant that only an intellectual could ignore or evade it."
—Thomas Sowell

In the 1990s, a group of scholars led by French scholar Stéphane Courtois documented the total communist death toll in a tome called *The Black Book of Communism*. The devil is truly in the details. *The Black Book* estimates that between 85 million and 100 million human beings lost their lives to communist experiments in the twentieth century.[9] Never has an idea had such catastrophic consequences. It illustrated a grim, simple equation: extreme moral passion minus reality equals mass death.

TABLE 1. DEATHS BY COMMUNIST REGIMES IN THE TWENTIETH CENTURY: A ROUGH TALLY

China	65 million
U.S.S.R.	20 million
North Korea	2 million
Cambodia	2 million
Africa	1.7 million
Afghanistan	1.5 million
Vietnam	1 million
Eastern Europe	1 million
Latin America	150,000
International Communist movement	about 10,000[10]

Source: The Black Book of Communism (Cambridge, MA: Harvard Univ. Press, 1999), 4.

WAS THE EARLY CHURCH COMMUNIST?

As a worldwide movement, communism is dead, even as it lingers in places like North Korea, Havana, and Harvard. Oh, there's a Communist Party USA, but it claims a membership of only fifteen thousand. Even if you combine them with the less radical socialist groups, however, you scarcely get 1 percent of the population, about the same percentage who have PhDs. (I assume that's a coincidence.) So am I beating a dead horse?

Few Christians, including Christian critics of capitalism, would now endorse communism. But what about the early church? Wasn't it communist? Here's how the book of Acts describes the first church in Jerusalem, which formed after the Holy Spirit descended upon the first Christians at Pentecost:

> Now the whole group of those who believed were of one heart and soul, and no one claimed private ownership of any possessions, but everything they owned was held in common. . . . There was not a needy person among them, for as many as owned lands or houses sold them and brought the proceeds of what was sold. They laid it at the apostles' feet, and it was distributed to each as any had need. (Acts 4:32–35)

Many who have read this passage have wondered if the Christian ideal isn't communism. After all, this was the first church in Jerusalem. They were "filled with the Holy Spirit and spoke the word of God boldly" (Acts 4:31). If they didn't get it right, who did?

On the surface, this looks like communism. But it's not. First of all, unlike modern communism, there's no talk of class warfare here, nor is there any hint that private property is immoral. These Christians are selling their possessions and *sharing* freely and spontaneously. Second, the state is nowhere in sight. No government is confiscating property and collectivizing industry. No one is being coerced. The church in Jerusalem was just that—the church, not the state. The church doesn't act like the modern communist state. No one in Acts gets their stuff confiscated. As Ron Sider notes, "Sharing was voluntary, not compulsory."[10] Third, when Peter later condemns Ananias and Sapphira for keeping back some of the money they get from selling their land, he condemns them not for keeping part of the proceeds of the sale, but for lying about it:

> Ananias . . . why has Satan filled your heart to lie to the Holy Spirit and to keep back part of the proceeds of the lands? While it remained unsold, did it not remain your own? And after it was sold, were not the proceeds at your disposal? How is it that you have contrived this deed in your heart? You did not lie to us but to God! (Acts 5:3–4)

Peter takes for granted that the property was rightfully theirs, even after it was sold.

Fourth, the communal life of the early church in Jerusalem is never made the norm for all Christians everywhere. In fact, it's not even described as the norm for the Jerusalem church. What Acts is describing is an unusual moment in the life of the early church, when the church was still relatively small. Also, many of the new Christians probably had come from a long distance to worship in Jerusalem at Pentecost. These new Christians would have had to return home soon after their conversion had it not been for the extreme measures taken by the newborn church to allow these Christians to stay and be properly trained in discipleship.

Compared with modern nation-states, the Jerusalem church was a small community banding together against an otherwise hostile culture. The circumstances were peculiar. For all we know, this communal stage lasted six months before the church got too large. Paul elsewhere told the Thessalonian Christians to "earn their own living" and sternly warned that "anyone unwilling to work should not eat" (2 Thess. 3:10, 12). So it's no surprise that the early communal life in Jerusalem was never held up as a model for how the entire church should order its life, let alone used to justify the state confiscating private property.

There have been Christian communities throughout history that have tried to live communally, and many monasteries and religious orders are more or less communal to this day. The ones that survive are small and voluntary. The others fall apart.[11]

THOSE PILGRIM COMMUNISTS

In fact, even voluntary communist experiments usually fail. American children hear the story of William Bradford at Thanksgiving. Bradford was the architect of the Mayflower Compact and the leader of a band of separatists who founded the Plymouth Colony in Massachusetts in 1620. Most kids learn that the colony lost half its population during its first, harsh winter. But few know about the colony's brief, and tragic, experiment with a type of communism.

Because of a deal made with the investors who funded the expedition, the pilgrims held their farmland communally rather than as individual plots. Food, work, and provisions were then divided evenly among the colonists. Sounds nice; but before long, conflicts arose among the colonists. In his journal Bradford reports what economists and common sense predict. In large groups, such an arrangement leads to perverse incentives, in which the lazier members take advantage of the harder working. They become "free riders" on the system. The other members grow more and more frustrated, and less and less productive. That's exactly what happened in the early years of the Plymouth Colony.

To solve the problem, Bradford soon decided to divide the plots among the individual families. Suddenly people had strong incentives to produce, and they did. Over the years, more and more of the land was privatized, and the colony eventually became a prosperous part of the Commonwealth of Massachusetts.[12] If Bradford had not had the guts to divide the commune into private lots, our schoolchildren would not be making little cutouts of turkeys and *Mayflower*s every November, since there probably would have been few if any survivors.

TALKING THE TALK

Recently, I was in Los Angeles and happened to see a teenage girl wearing a Che Guevara T-shirt in hip, revolutionary colors: red and black. We were waiting in line for a latte, so I asked the girl if she knew who Che Guevara was. She was a little sketchy on the details, but she knew he was a voice for the weak and the powerless. She mumbled something about his standing up to the Establishment and defending the poor. And anyway, he looked really cool with his beret and rugged good looks.

I've seen a lot of these T-shirts, usually on teenagers born after the collapse of the Soviet Union. Most probably share the views of this young Californian. They all need to read up on the subject. Guevara was an Argentinean communist revolutionary who trained Fidel Castro and who oversaw the first execution squads and created the labor camps in Communist Cuba. He was executed in Bolivia in 1967 after trying to overthrow the government with terrorist tactics. He was a bloodthirsty thug.[13] What's the attraction?

It's this: communists like Che Guevara had their rhetoric right. They knew the value of good PR. They talked a good talk, denouncing inequality and defending the poor. Despite the nasty outcome of their experiments, they can still get a pass from those who sympathize with the ideals these men presented to the larger world.

When the Soviet Union collapsed, the Moscow correspondent for one Western paper addressed the Russians wistfully: "Thanks for having tried."[14] We tend to admire someone who tries and fails miserably over someone who does nothing at all. That's why any sensitive person who wants to *do* something, anything, to fight injustice can be drawn to revolutionary rhetoric. This isn't all bad. In the book of Revelation, Jesus tells the church of Laodicea, which took pride in its wealth, "I know your deeds, that you are neither hot nor cold. I wish you were either one or the other! So, because you are lukewarm—neither hot nor cold, I am about to spit you out of my mouth" (Rev. 3:15–16).

Strong words. God doesn't like apathy. Neither do serious Christians. This is why revolutionary rhetoric can be tempting. Revolutionaries are anything but apathetic. Their rhetoric is coursing with passion, and it can capture the mind. For a time, it captured mine. Despite communist brutalities, I still wanted to build a just society, which for me meant equality. I still liked the goals, if not the real outcomes.

Who is not moved by Martin Luther King Jr.'s letter from a Birmingham jail, in which he points out that you can be a thermometer and make a record, or a thermostat and correct the wrong? Surely we should be thermostats. James tells us that faith without works is dead. Our faith is "brought to completion" by our works (James 2:17, 22). Some might be reminded of Marx's famous dictum: "The philosophers have only interpreted the world in various ways: the point, however, is to change it."

That's a great quote for a dorm-room wall; but what does it mean to change the world? Some things can be changed, like the wallpaper in your bathroom. Other things can't be changed, like gravity. And some things shouldn't be changed, like the love of a mother for her children. Before trying to experiment with half mankind,[15] it might have been a good idea for communists to ask a few questions, like: What is man really like? What does the state have a right to change? What is within its power to change? What is beyond its control? And then they should have made sure that the changes they made were for the better, not the worse.

Instead, the communists sought to create a society in which everyone was equal, to establish a "heaven on earth," to attain nirvana as they imagined it. They plowed ahead without asking key questions about reality. Some, such as Mao, seemed to deny a reality apart from our minds, to which we must adapt, preferring instead to try to create their own reality by the sheer power of mind and will.[16] Others, like Lenin, adjusted Marx's system because they discovered to their dismay that it didn't square with reality. The history of communism in the last century is the starkest example of what happens when you tangle with reality: you lose, and so do the subjects of your experiment.

"Socialism only works in heaven, where they don't need
it, and in hell where they already have it."
                              —attributed to Ronald
                              Reagan

Communism failed because it treated human beings and
human society like wet clay. "It is on a blank page," wrote Mao,
"that the most beautiful poems are written."[17] Instead of starting
with man as he really is, communists sought to create their ideal
man in an ideal society, which for them meant one in which perfect equality reigned. To have any chance of success, they had to
abolish private property, since property is the most obvious sign
of inequality. That doesn't mean no one controls the property. It
means the state must confiscate and control whatever property is
in private hands. So the state simply must coerce and kill, since
most people don't willingly hand over their property to aggressors. The communists tried to draw heaven down to earth. They
brought up hell instead.

Marx spoke naively of "changing the world." There's much
more wisdom in Reinhold Niebuhr's "serenity prayer," repeated
by millions of alcoholics over the years:

Lord, grant me the serenity to accept the things I cannot
change, the courage to change the things I can, and the
wisdom to know the difference.

ALREADY BUT NOT YET

Christians believe that the world not only can but will be changed
for the better when Christ comes fully into his kingdom. But
when's that, and what good does it do us here and now? The
Bible describes the present as an in-between time. God created the
world good, but we have fallen. Things aren't the way they're supposed to be. Each of us bears the effects of sin. No one is immune
to those effects. "If only there were evil people somewhere insidi-
ously committing evil deeds," said Aleksandr Solzhenitsyn, "and

it were necessary only to separate them from the rest of us and destroy them. But the line dividing good and evil cuts through the heart of every human being." No one is fit to be a benevolent dictator, since, as Lord Acton said, "power tends to corrupt; absolute power corrupts absolutely."

Even the wider creation bears sin's effects. "The creation was subjected to futility . . . ," said Paul to the church at Rome, "in hope that the creation itself will be set free from its bondage to decay and will obtain the freedom of the glory of the children of God" (Rom. 8:20–21).

That's the bad news. The good news is that God will establish a kingdom of peace and justice—a new heaven and a new earth—in which evil and death will be vanquished forever. So no matter how dire our present condition, the Christian has hope.

At the same time, God doesn't reserve all the promises of his kingdom for the future. He established his church, after all. Jesus said the kingdom of God is like leaven that works its way through the whole lump. We shouldn't look for heaven on earth, but we should expect salt and light and leaven. And as Christians, we should expect to find ourselves as part of the leavening process.

To grasp the biblical message of God's reign, we must avoid two tempting but false extremes. The first temptation is to quarantine God's kingdom safely in the distant future up in the clouds, in "heaven," sealed off away from the blood, sweat, and tears of the present. In this view, we should expect the world to be not only corrupt but beyond repair. There's nothing we can do about it. The world's going to hell in a handbasket. Don't bother polishing the brass on a sinking ship. The best we can hope for is that in the end we'll be saved, maybe raptured before it gets really bad, and perhaps we'll be able to bring a few converts along with us. In this telling, Christian faith is at worst a story about me-and-Jesus, about saving my soul and little else, and at best it's about a gospel message that can save souls but has little power to transform the larger world for good. On this model, God's kingdom has little to do with worries about poverty, injustice, and the physical struggles that mark our earthly lives.

   This is not how the Bible describes the kingdom of God. In the New Testament, the kingdom of God is described as in some way a present reality. In preaching repentance before Jesus began his ministry, John the Baptist said, "Repent, for the kingdom of God is at hand" (Matt. 3:2). Jesus told his disciples to tell others, "The kingdom of heaven is near" (Matt. 10:7). The Jewish people were expecting their Messiah (their "anointed one") to appear and deliver them from their oppressors—in this case, from the Romans and their collaborators. This wasn't some airy fairy spiritual kingdom, but rather an earthly reign of justice and peace in which Israel would again be a great nation. Jesus came to fulfill the Old Testament prophecies, not to dash them. And yet he didn't do what the Jews of his day were expecting.

   Unlike an ordinary kingdom, his kingdom, Jesus said, is "not of this world" (John 18:36). He told Nicodemus: "I tell you the truth, no one can see the kingdom of God unless he is born again" (John 3:3). When asked by Pharisees when the kingdom of God would come, he told them that "the kingdom of God is among you" (Luke 17:21). In the present at least, the kingdom of God is something we must receive as a small child or not at all (Mark 10:15).

   But just after Jesus told the Pharisees that the kingdom was already present, he turned and told his disciples that it would come suddenly in power and judgment, when no one expects it (Luke 17:21ff.). The New Testament describes a cataclysmic future time in which God's kingdom will come in power for everyone to see.[18] God will destroy everything that opposes his reign (1 Cor. 15:24–27). There will be a new heaven and a new earth, a New Jerusalem, in which God will make everything new, will wipe away all tears, and will defeat death (Rev. 21:1–5).

   This is clearly in the future, since, contrary to everyone's expectations, Jesus didn't take power in Jerusalem as a king, but humbled himself, and allowed himself to be crucified. His resurrection from the dead is the firstfruit of the general resurrection in the kingdom to come, in which every knee shall bow and every tongue confess that he is Lord (1 Cor. 15:23; Phil. 2:9–11).

This doesn't contradict the Old Testament picture of God's kingdom. The Old Testament prophets also spoke of a messianic kingdom at the end of time. Isaiah prophesied:

The wolf will live with the lamb,
the leopard will lie down with the goat,
the calf and the lion and the yearling together;
and a little child will lead them. (Isa. 11:6)

So Christ will consummate his kingdom in power at the end, even though it has already come like a seed that quickly grows underground before springing up (Mark 4:26–29), and like yeast, which, when mixed with flour, slowly leavens all of it (Matt. 13:33). The kingdom breaks into the present as God broke into history in Jesus. We should expect signs of God's reign even in our age, when death and sin are still very much with us. Wherever believers are obeying God's commands in the world, there should be some glimmer of his kingdom. *That includes our economic and political life.* We aren't called to sit on our hands and wait for the Lord to return. Remember Kuyper's statement: "There is not a square inch in the whole domain of our human existence over which Christ, who is Sovereign over all, does not cry: 'Mine!'" We should be about the business of his kingdom here and now.

That doesn't mean, however, that *we* will establish God's kingdom in its fullness through our own good works. God is responsible for establishing his kingdom, not us. God came first in humility in Jesus. He will come again in power and glory. But that hasn't happened yet, and we can't trigger it. If we try, we won't just fail, we'll do far more harm than good.

The grand communist experiment is a secularized attempt to establish God's kingdom on earth. Marx's story has the main elements of the Christian story: primeval paradise, fall, redemption, eternal paradise. It's just stripped of references to God, sin, Jesus, and the afterlife. If Christians can't bring about the kingdom of God on earth, however, it's no surprise that this secular surrogate was doomed to failure as well.

To believe otherwise is to believe the Nirvana Myth.[19] This is the second temptation. The Nirvana Myth is not simply the belief that good will triumph in the end or the belief that the kingdom of God is already present in history. It's the delusion that we can build utopia if we try hard enough, and that every real society is intolerably wicked because it doesn't measure up to utopia.

Myth no. 1: The Nirvana Myth (contrasting capitalism with an unrealizable ideal rather than with its live alternatives)

We learned in the twentieth century that acting on this myth can be disastrous. Never has there been a greater gap between ideals and outcomes than in communism. In fact, so many people would not have been led astray if communism had advertised baser goals. No, communist brutalities needed the cover of some grand moral vision. Communism appealed to, even if it inverted, man's moral impulse. This is the worst outcome of the Nirvana Myth.

But the myth can have subtle effects even if we reject utopian schemes. To avoid its dangers, we have to resist the temptation to compare our live options with an ideal that we can never realize. When we ask whether we can build a just society, we need to keep the question nailed to solid ground: just *compared with what?* It doesn't do anyone any good to tear down a society that is "unjust" compared with the kingdom of God if that society is *more* just than any of the ones that will replace it.

Compared with Nirvana, no real society looks good. Compared with utopia, Stalinist Russia and America at its best will both get bad reviews. The differences between them may seem trivial compared to utopia. That's one of the grave dangers of utopian thinking: it blinds us to the important differences among the various ways of ordering society. The Nirvana Myth dazzles the eyes, to the point that the real alternatives all seem like dull and barely distinguishable shades of gray. The free exchange of wages for work in the marketplace starts to look like slavery. Tough competition for market share between companies is confused with theft and survival of the fittest. Banking is confused with usury and exploitation. This shouldn't surprise us. Of course a modern capitalist society like the United States looks terrible compared with the kingdom of God. But that's bad moral reasoning. The question isn't whether capitalism measures up to the kingdom of God. The question is whether there's a better alternative in this life.

> "Those who condemn the immorality of liberal capitalism do so in comparison with a society of saints that has never existed—and never will."
> —Martin Wolf, *Why
> Globalization Works*

If we're going to compare modern capitalism with an extreme, we should compare it with a *real* extreme—like communism in Cambodia, China, or the Soviet Union. Unlike Nirvana, these experiments are well within our power to bring about. They all reveal the terrible cost of trying to create a society in which everyone is economically equal.

If we insist on comparing live options with live options, modern capitalism could hardly be more different, more just, or more desirable than such an outcome. That doesn't mean we should rest on our laurels. It means we need to stay focused on reality rather than romantic ideals.

So how should we answer the question that began this chapter: can't we build a just society? The answer: we should do everything we can to build a *more* just society and a more just world. And the worst way to do that is to try to create an egalitarian utopia.

Two

What Would Jesus Do?

During my senior year in college, I decided to read the Bible straight through quickly rather than in bite-sized chunks. When I did so, a larger pattern jumped out at me: God's abiding concern for the poor, and his expectation that we share his concern. That's the message from Genesis to Revelation.

Near the beginning, God tells the Hebrews before they enter the Promised Land not to "be hard-hearted or tight-fisted toward your needy neighbor. . . . Since there will never cease to be some in need on the earth, I therefore command you, 'Open your hand to the poor and needy neighbor in your land'" (Deut. 15:7, 11). The Proverbs are chock-full of commands that connect our love of God with how we treat the poor: "Those who oppress the poor insult their Maker / but those who are kind to the needy honor him" (Prov. 14:29).

Outraged Old Testament prophets like Amos announce God's judgment on Israel for defrauding those who can ill afford it, trampling the needy, and "bringing to ruin the poor of the land" (Amos 8:4–6). No minced words here: God let the Babylonians and Assyrians carry off the Jews into captivity because the Jews worshipped false gods and failed to care for the poor and needy.

It's the same story in the New Testament. God became flesh and was born to a humble woman in a stable filled with animals. Poor shepherds were the first to hear the news and to see God become man. And Luke tells us that at the beginning of Jesus's ministry, Jesus entered the synagogue in Nazareth—his hometown—and read from the prophet Isaiah:

The Spirit of the Lord is upon me,
because he has anointed me
to bring good news to the poor.
He has sent me to proclaim release to the captives
and recovery of sight to the blind,
to let the oppressed go free,
to proclaim the year of the Lord's favor. (Luke 4:18–19)

When Jesus told the listeners that he was there to fulfill Isaiah's prophecy, they got so mad that he had to move on to Capernaum.

Jesus's parable of the sheep and the goats is really hard-core, and troubles the placid doctrine of salvation by faith alone. Jesus says that when the Son of Man returns to establish his kingdom, he will separate the "sheep" and "goats" depending on how they have treated people who are hungry, thirsty, strangers, naked, sick, and imprisoned. The goats are dispatched to "the eternal fire prepared for the devil and his angels." The sheep, in contrast, are told to "inherit the kingdom prepared for you from the foundation of the world." That sounds much nicer, especially without all the goats. In the parable, the only difference between the sheep and the goats is how they treated the vulnerable, and so, by extension, how they treated Jesus himself: "Truly I tell you, just as you did it to one of the least of these who are my family, you did it to me" (Matt. 25:31–46).

At one point, a religious figure asks Jesus, "Teacher, which commandment in the law is the greatest?" Jesus says, "'You shall love the Lord your God with all your heart, and with all your soul, and with all your mind.' This is the greatest and first commandment. And a second is like it: 'You shall love your neighbor as yourself.' On these two commandments hang all the law and the prophets" (Matt. 22:34–40).

Another time a lawyer asks Jesus how he can inherit eternal life. Jesus tells him he has to fulfill these two greatest commandments: love God to the full, and love your neighbor as yourself. The lawyer, sensing an unattainable goal, looks for a semantic loophole in the word *neighbor*. "And who is my neighbor?" he asks. Jesus then tells the story of the Good Samaritan. If you read it carefully, you'll notice that Jesus never answers the lawyer's question. Instead, after he tells the parable, he asks a different question: "Which of these three, do you think, was a neighbor to the man who fell into the hands of the robbers?" (Luke 10:25–37). His point seems to be that instead of trying to whittle away the definition of *neighbor* to get off easy, we should strive to *be* the good neighbor to everyone else. Jesus's reversal prevents the man from simply picking a safe neighborhood where none of his "neighbors" will ever need help. Even if he lives in a gated community, his neighborhood extends well beyond the gate.

So God's concern for the poor isn't some sidelight. It follows straight from what Jesus tells us are the two greatest commandments. For any follower of Jesus, then, that we should care for and help the poor is not the question. The question is, How do we do it? "Piety," said the Christian philosopher Etienne Gilson, "is no substitute for technique." What he meant is that having the right intentions, being oriented in the right way, doesn't take the place of doing things right. A pilot's caring deeply for his passengers and wanting to land a plane safely are no substitute for his learning how to actually land planes safely. Jesus suggests the same thing. "Love the Lord your God with all your heart and with all your soul and with all your mind." Don't forget the third item: love the Lord with all your *mind*. And don't misunderstand *heart*. Your heart isn't just your feelings. In the Bible, *heart* refers to the seat of your will and your emotions. I hope you already have a heart for the poor. Lots of Christians do. But do you have a *mind* for the poor? Unfortunately, that's in rather short supply.

God knows your heart. Spiritually you're better off a little mixed up about economics than indifferent to human suffering. Economically, though, only *what* you do is important, whatever your reason. Buying a bunch of bananas at Costco will have the same economic effect no matter why you buy them.

Our minds and our motives aren't isolated compartments. God gave us minds and reason, so we're responsible for thinking through the consequences of our actions. In fact, it's morally self-indulgent to feel good about our motives when it comes to actions that affect the world. When the focus is personal spiritual growth—sure, we should examine our motives. But when it's time to roll up our sleeves and actually try to help someone, fixating on our motives can become a stumbling block. It can distract us from discovering the right action at the right time. Teenagers rightly ask, "What would Jesus do?"

I'm going to sound like a schoolmarm, but I'll say it anyway: we need to exercise *prudence*. Prudence means "to see reality as it is, and to act accordingly," to conform your mind, and then your actions, to reality.[1] It's one of the four cardinal virtues, along with justice, fortitude, and temperance. *Cardinal* is based on the Latin word *cardo,* which means "hinge." These four virtues are cardinal because all the other virtues hinge on them. Helping the poor, for instance, hinges on prudence. That's because, in the economic realm, actions have all sorts of unintended consequences. We can't anticipate all of them. But we can anticipate a lot of them. Therefore, if we really want to help the poor, we have to exercise prudence—to know what the world is really like, and act accordingly.

Henry Hazlitt, an economic journalist, thought this was so important that he defined economics in terms of consequences. "The art of economics," he said, "consists in looking not merely at the immediate but at the longer effects of any act or policy; it consists in tracing the consequences of that policy not merely for one group but for all groups."[2]

This is one of those principles that are easy to get and even easier to forget. Unfortunately, Christians have supported all sorts of policies that were well motivated but that made matters worse, not better. Let's look at a few popular policies that are long on compassion but short on prudence.

> "The art of economics consists in looking not merely at the immediate but at the longer effects of any act or policy; it consists in tracing the consequences of that policy not merely for one group but for all groups."
> —Henry Hazlitt

A "LIVING WAGE"

The "living wage" sounds nice. Who wouldn't want everyone to make enough money so that they can live well? The problem is that "living wage" policies create some nasty unintended consequences. Those who want every household to have a living wage usually try to achieve this through minimum-wage laws, where the government sets the minimum amount an employer may pay an employee. Such laws replace cooperation with coercion, since they make it a crime for individual employers and employees to enter freely into agreements for work and wages. This free market is replaced by somebody in Congress deciding what's best for all concerned. That alone is troubling. Worse, the policy doesn't even do what it's supposed to do—help the poor.

In the summer of 2007, President Bush signed a bill raising the federal minimum wage from $5.15 to $5.85 an hour. In the summer of 2008, it went to $6.55. And in July 2009, it will go to $7.25. These events mark the first time the federal minimum wage had been raised in ten years. When the Senate passed the bill, evangelical activist Jim Wallis appeared at a press conference with Senator Ted Kennedy and others. He said that "God hates inequality," and that the bill was "only the beginning." To establish the point, he asked: "What does the Bible have to say about the minimum wage?" His answer: "The prophet Isaiah said: 'my chosen shall long enjoy the work of their hands. They shall not labor in vain . . .'" (65:22–23). Huh? God is saying that his people will not have what they have earned or reaped stolen or destroyed, not that the government will come in and artificially jack up the wages paid to day laborers. The verse from Isaiah is part of a prophecy about the "new heavens" and "new earth" that God (not Congress) will someday establish. A few verses later, the text describes this as a time when the "wolf and the lamb shall feed together" (65:25). I don't think that's happening yet.

Wallis also quoted James: "James, the sibling of Jesus [who] probably knew what his brother thought about things pretty well: 'Listen! The wages of the laborers who mowed the fields, which you have kept back by fraud, cry out, and the cries of the harvesters have reached the ears of the Lord.'"[3] But this passage clearly refers to workers who haven't been paid what they were promised. If someone agrees to pay a kid five dollars to shovel his walk and then when the kid's finished tells him, "I was joking about paying you," he's just robbed the kid of five dollars. That's the sort of unjust behavior James is describing.

So neither of these verses has a thing to do with a federal minimum wage. Wallis's argument is mere demagoguery. Demagoguery doesn't become prophecy just by misquoting a Bible verse. Unfortunately, in politics, "demagoguery beats data," as former House Majority Leader Dick Armey put it. Nevertheless, Christians who prize truth should tie their policy ideas to data, not demagoguery. So what is the data on minimum wages?

Let's think about it. What would happen if the minimum wage were a thousand dollars an hour? Wouldn't that be great? Even the gal cleaning restrooms down at the mall would pull in forty thousand dollars a week. Problem is, five minutes after the law went into effect, nobody would be getting paid (legally) to clean the restrooms at the mall. And the poorly educated woman with a good work ethic would be out of a job, along with a flood of other people. "You cannot make a man worth a given amount by making it illegal for anyone to offer him anything less," writes Hazlitt.[4] To a business, employee wages are costs. The fact that Washington sets a minimum wage doesn't mean an employer can pay it. That's why countries with high minimum wages like France tend to have higher unemployment than countries with low minimum wages (like the United States). That's no surprise. A wage is a price on a commodity—labor. Different kinds of labor, such as dishwashing and retina surgery, are going to have different values economically, depending on the who, where, when, what, and how of the labor. A minimum wage fixed by law ignores that reality. It's a form of price fixing that tries to distribute wealth before it's been created.

Of course, it's harder to trace the consequences with less extreme minimum-wage laws. If most people already make more than the minimum wage (as they do in the United States), a forced wage hike seems like no big deal. It may lead just to slightly higher prices spread out, so no one connects the dots. Moreover, some people will benefit, at least on the surface. (This is why few politicians have the guts to oppose minimum-wage laws directly.) Union employees, for instance, who rarely make minimum wage, like these laws since they make it more costly for employers to hire less-skilled workers who have to be trained.[5]

The people most likely to suffer from these laws are those at the bottom: the unskilled, young, inarticulate, and handicapped workers, who need to grab the bottom rung of the economic ladder. Raise it too high, and the rung is out of reach. In a diverse economy, remember, low-paying jobs are *entry-level jobs*. Like it or not, some people need entry-level jobs very close to ground floor. Few people stay in these jobs forever. The experience and connections that such jobs provide can be more valuable than the salary itself. Minimum-wage laws favor vocal and visible workers over the vulnerable workers who can least afford to be unemployed. Good intentions don't change that.

FAIR TRADE

You've probably bought something with a fair-trade label on it. I have. If you buy a latte at Starbucks, for instance, you have the option of buying more expensive "fair trade" coffee. Who wouldn't want fair trade? "No, thanks, I'll take a cup of unfair-trade coffee." For a Christian it sounds like a no-brainer. But what exactly is "fair trade" coffee?

With fair-trade coffee, the coffee farmers are paid twice the market price (around $1.26 per pound as of this writing) or more, based on an estimate of how much they need to enjoy a decent standard of living. That higher price, plus the costs of monitoring a "fair trade" supply chain, make the coffee more expensive at Starbucks. But lots of people feel good about paying an extra fifty cents for fair-trade joe, and lots of religious organizations like the Interfaith Fair Trade Initiative and the Presbyterian Coffee Project are getting religious folks on board.[6] The church I attended when I was living in Seattle provided only fair-trade coffee.

The coffee plant is fussy. It needs just the right place and climate to grow. Most of the coffee-plant-friendly places are in third-world countries rife with corruption, lame property laws, and wobbly rule of law. From my unscientific polling while sipping my double tall nonfat sugar-free cinnamon dolce lattes with a small dollop of whipped cream, I've learned that lots of people are vaguely aware of these problems. Therefore, they don't like the idea that coffee pickers in South America are making, say, ten cents to pick the coffee beans that Starbucks uses for a double latte that costs $4.28. That's a sweet deal for Starbucks, but where's the justice in that?

There's plenty of injustice in the world, but that price difference isn't proof of it. It's easy to forget that the economic value of the coffee has changed from Peru to Portland. A pound of green coffee beans in Brazil hasn't been packaged, preserved, labeled, shipped, delivered, carefully roasted, prepared to extremely high maintenance specifications with expensive equipment, and served piping hot to a customer in a nicely appointed Starbucks in Bellevue, Washington. Economic value has been repeatedly added to the beans as they go up the supply chain. As objects of value, the beans in Brazil and the breve in Bellevue are two different things. As long as the Brazilian coffee pickers are free workers, the different prices may simply reflect the transformation. That's not unfair. In fact, the coffee farmers are making as much as they are because of the huge demand for coffee beans created, in part, by companies like Starbucks.[7] They're not as well off as Starbucks execs by a long shot, but they're better off than they would be without a high-end espresso market in the United States.

While some so-called fair traders seek to have government coercively set a minimum price for all coffee beans, at the moment nobody is *making* customers buy fair-trade coffee.[8] They're buying it freely, so they must see some value in it. Perhaps it's just the good feeling of knowing they're helping poor coffee farmers. There's a market for that feeling; otherwise, Starbucks couldn't sell the coffee. So what's the problem with buying fair-trade products, as long as it's voluntary? Isn't it just a market-oriented way to deliver charity?[9]

The problem is subtle. Paying artificially high prices for some coffee encourages poor farmers to enter or stay in the coffee market when it's against their long-term interest to do so. Consider this statement by one fair-trade organization, Global Exchange:

> Coffee prices have plummeted and are currently around $.60–$.70 per pound. "With world market prices as low as they are right now, we see that a lot of farmers cannot maintain their families and their land anymore. We need Fair Trade now more than ever," says Jerónimo Bollen, Director of Manos Campesinas, a Fair Trade coffee cooperative in Guatemala.

There's a reason the market prices have dropped. In recent years, millions of people have started drinking different kinds of high-end coffee, so more farmers and companies have entered the market around the world. Vietnam is now a major coffee exporter.[10] When the supply goes up, the price for coffee goes down—not because of injustice, but because of the law of supply and demand. Some farmers who were competitive in 1990, however, are no longer competitive. There's no law of economics or morality that sets the price of coffee high enough so that every coffee farmer everywhere will always be able to make a decent living growing coffee—anymore than there's a law that everyone will always be able to make a decent living manufacturing tallow candles or buggy whips or eight-track tapes or Winnebagos. Markets are like the weather in Michigan: they change whether we like it or not.

While the market price for raw coffee will get more competitive, the artificially high "fair trade" prices are encouraging some farmers to enter and stay in quirky markets when their products are not actually competitive. A market price is a sign of an underlying reality (more on this later). A dropping price tells producers and sellers that supply is exceeding demand;

they can use that information to adjust and reallocate scarce resources like time, land, and labor to more valued uses. A rising price signals the opposite. Farmers in fair-trade schemes are deprived of this information. It's in the long-term interest of some of these farmers to start growing more-competitive products or even to move out of agriculture altogether. Moreover, the small percentage of farmers in fair-trade schemes are being favored arbitrarily over farmers in most places, who don't have access to such a scheme.

As it is, the future livelihood of farmers now benefiting from "fair trade" schemes depends on millions of people continuing to pay high "fair trade" prices while remaining clueless of the economics. That's a dangerous gamble, since "fair trade" prices will rise with inflation, while the normal market price could stay the same or go down. (It's at historic lows now.) This will shrink the demand for fair-trade coffee. When the price gets high enough, many pious fair-trade coffee drinkers will switch to regular coffee, or start drinking chai lattes instead. Then there will be way too many coffee farmers producing way too much "fair trade" coffee, which will devalue it even more. How's that fair?

Even if a few people *would* pay for a fair-trade label at any price, their charity would be better channeled elsewhere. For instance, most of the countries that provide coffee have abysmal property laws. Many farmers don't have solid titles to their land, so they're neither willing nor able to make a long-term investment in it or to sell it and do something else. They're stuck. Even those who have titles often need advice on what to produce. Many organizations are working to improve these situations.[11] All the money being wasted creating perverse incentives in the coffee market and other markets could instead support projects that do long-term good for third-world farmers, even if those projects don't provide a quick compassion fix for American coffee drinkers.

Myth no. 2: The Piety Myth (focusing on our good intentions rather than on the unintended consequences of our actions)

FOREIGN AID

Rich countries give all sorts of things to poor countries and call it "aid." Sometimes we dump cotton in places like Senegal. In the United States we overproduce cotton because of subsidies that inflate the price. We end up with too much cotton, and we've got to do something with it. So we dump it on poor countries and call it "aid." It sounds nice, but all that extra cotton suppresses cotton farming in these poor countries, which could do much better if they could compete fairly in a free world market for cotton. Such policies are not "aid," despite the nifty ad slogans.

But the United States and other rich countries also send billions of dollars in aid for development through organizations like the United Nations, the World Bank, and the International Monetary Fund. These organizations' real purpose is just what is advertised: to alleviate third-world poverty. Economist Paul Collier, who spent years at the World Bank, describes aid for development this way: "We used to be that poor once. It took us two hundred years to get to where we are. Let's try to speed things up for these countries."[12] This is admirable clarity. Now that we know how countries become wealthy, surely we could help speed up the process for those still far behind.

Think about technology. It took us a century and a half to go from the pony express to telegraph and telephone lines to copper wire to cable, fiber optics, cell phones, and satellites. But rural Botswana need not slog through the same slow process to get cell-phone coverage. And if we can speed up the adoption of new technology, why couldn't we speed up the methods for creating wealth?

Imagine the route to wealth creation as a path up a treacherous mountain like Mount Everest. Scores of people summit Everest every year. But we remember Sir Edmund Hillary and his expedition, because they were the first to do it, in 1953. Being first is a big deal, because the first summiting of a mountain is the most dangerous, since the climber doesn't know the best way up. Many climbers may fall to their deaths or freeze in hidden crevasses before anyone finds a safe passage up. Once someone has reached the peak and returned safely to the base, however, future climbers can follow the same path, usually without dying. At its best, that's what aid is supposed to do: help poor countries develop more quickly by encouraging them to take the known pathways for creating wealth rather than wandering over economic cliffs.

Recent efforts like the ONE Campaign ("The Campaign to Make Poverty History") have made millions of people aware of the plight of the poor in third-world countries. With tireless promotion by stars like Bono, the lead singer of the band U2, the ONE Campaign has become the compassion fashion of college students everywhere. The ONE Campaign doesn't just give out little white wristbands. It has several laudable goals related to the United Nations' Millennium Development Goals:

- Reduce by half the number of people in the world who suffer from hunger

- Provide free access to primary education for 77 million out-of-school children

- Provide access to clean water for 450 million people and to basic sanitation for 700 million people

- Prevent 5.4 million young children from dying of poverty-related illnesses each year

- Save sixteen thousand lives a day by fighting HIV/AIDS, tuberculosis, and malaria[13]

As *Christianity Today*'s David Neff has observed, these goals are "all good."[14] So how do we accomplish them? The ONE

Campaign calls for people to lobby the U.S. Congress to increase foreign aid to 1 percent of the federal budget. (That's why it's called the ONE Campaign.)

In 2005, rock musician Bob Geldof organized a concert series and campaign, Live8, to encourage wealthy G8 nations to vote on debt relief and government aid to poor African nations. Such problems were the result of failed aid projects. Everybody from Tony Blair to Bono seemed to agree with Geldof. Some economists, doing their best to maintain their field's reputation as the "dismal science," pointed out that such debt relief might send the wrong message. It could encourage corrupt and wasteful governments to continue in their ways, reward countries with high debt while doing nothing for poor countries with little or no debt, discourage future credit and private charity, and generally mess up accountability for bad behavior.

Geldof's response to the naysayers was telling. "Something must be done," he said, "even if it doesn't work."[15] Geldof meant well, but he wasn't thinking well. The two options he considered—useless help or useful help—aren't jointly exhaustive. Almost anything we do can have one of *three* outcomes. It might help, it might make no difference, or . . . it might *hurt*. Geldof forgot the third possibility. When it comes to foreign aid, unfortunately, *hurt* is usually what happens.

Economist William Easterly has marveled at the "amazing recycling ability of the aid industry." It never seems to take account of the lessons of history. Government-to-government foreign aid has been the tool of choice for the last four decades as a way for wealthy countries to lift poor countries out of poverty. Rich countries have sent $2.3 trillion to poor countries in the past fifty years. (Over half a trillion dollars went to Africa.)[16] While some aid has helped a few things around the edges, it has utterly failed as a solution to third-world poverty. More often than not, the money has either been wasted or helped prop up the corrupt dictators who keep their countries poor. In one sad case, a group of refugees had to flee from an African dictator who was supported by foreign aid. To balance the scales, the refugees were awarded aid as well. Guns and butter for everyone, just to be fair? This sounds like a solution in serious need of dissolution.

Bert once told Ernie on *Sesame Street*, "It's easy to have ideas, but it's not so easy to make them work." That's foreign aid. A recent study by the National Bureau of Economic Research showed *no* correlation between the amount of aid a country receives and its economic growth, even when you account for the fact that it's the poorer countries that normally receive aid. This doesn't prove that no form of aid could ever help any country. But it puts the lie to the easy assumption that government-to-government aid as currently practiced helps poor countries.[17]

Besides the evidence, there's a simple but forgotten question: what does tax money sent from one government to another have to do with speeding up the creation of wealth? Nothing. No developing country ever got rich that way.[18] Government-to-government aid isn't one of the basic ingredients for wealth creation. It might make us feel good to wear wristbands and sign petitions for the government to spend more of other people's money. But it's not going to do much for the poor in the developing world. The path to wealth is well known. As William Easterly puts it:

> The end of poverty will come as a result of homegrown political and economic reforms (which are already happening in many poor countries), not through outside aid. The biggest hope for the world's poor nations is not Bono, it is the citizens of poor nations themselves.[19]

That doesn't mean we can't help. It just means we should try something other than what has already been tried—and has already failed.

"Something must be done, even if it doesn't work."
—Bob Geldof

GOVERNMENT-RUN WELFARE

We all know about the welfare state. It's hard to imagine the world without it; but it wasn't always around. In the United States, it started with FDR's 1930s programs. They were meant to end the Great Depression but deepened and lengthened it instead.[20] Most of the federal programs around today, however, started with Lyndon Johnson's "War on Poverty," a part of his grandiose Great Society agenda implemented in the mid-1960s.

Judged by its results, however, the War on Poverty was more a War on the Poor. Statistically, the poverty rate was already on a steady decline before Johnson put his hand to the domestic plow. It had dropped from about 22 percent to 15 percent between 1959 and 1965. After 1965, just after the War on Poverty began, the poverty rate settled in, and it has continued to fluctuate between 12 and 15 percent up to the present day, even though American society is now much more prosperous overall.[21] So the least that can be said is that the War on Poverty cost trillions of dollars and didn't work. But the statistics hide the full cost: not only did the War on Poverty fail; it created economic and social problems worse than those it was meant to solve.

Although it's too soon to know what will happen with fair-trade coffee, most of us already know the unintended consequences of the welfare state: family breakdown and illegitimacy, especially among poor urban minorities and white Appalachians (in case you thought the problem was based on race); cycles of dependency that transfer from one generation to the next; anger, despair, and hopelessness. Activists often say this proves that our "society" is racist and unjust, but then why are these social problems so much worse now than they were in, say, the 1920s, when racism was much more rampant? In the twenties, for instance, there was little statistical difference in illegitimacy rates between blacks and whites. No, this is the poison fruit of a federal welfare system that was designed to help the disadvantaged but ended up rewarding destructive behavior.

As Aristotle said twenty-five hundred years ago, "If you want to encourage something, reward it. If you want to discourage it, punish it." If women are given more aid if they have more children with unknown fathers but cut off if a father is around, you can be pretty sure you'll end up with a lot of children without fathers and lot of women as wards of the state. And in prosperous societies like the United States, the best predictor of childhood poverty is being in a single-parent household headed by a mother.

The sad statistics of federal welfare are unmistakable:

- An experiment comparing a control group with household recipients of welfare benefits revealed that welfare is a disincentive to work. Husbands reduced their hours worked by an average of 9 percent, and wives reduced hours worked by 20 percent. Young males reduced their hours worked by 33 percent; singles, by 43 percent.

- The Seattle Income Maintenance Experiment and the Denver Income Maintenance Experiment concluded that every dollar of subsidy (guaranteed income supports) reduces labor and earnings by 80 cents.

- Data from the National Longitudinal Survey of Youth shows that a 10 percent increase in welfare benefits resulted in a 12 percent increase in out-of-wedlock births among low-income women aged fourteen to twenty-two.

- A study from the U.S. Department of Health and Human Services found that a 50 percent increase in the value of AFDC (Aid to Families with Dependent Children) and food-stamp payments led to a 43 percent increase in the number of out-of-wedlock births.

- When job programs began their massive expansion, the black youth unemployment rate began to rise. Between the years 1951 and 1980, black twenty- to twenty-four-year-olds experienced a 19 percent increase in unemployment. For eighteen- and nineteen-year-olds, the increase was a remarkable 72 percent.[22]

- Since 1970, out-of-wedlock birth rates have soared. In 1965, 24 percent of black infants and 3.1 percent of white infants were born to single mothers. By 1990, the rates had risen to 64 percent for black infants and 18 percent for whites. Every year about 1 million more children are born into fatherless families.[23]

Some of these problems have improved with the Welfare Reform Act signed by President Clinton in 1996. Unfortunately, despite all these facts, many Christian activists still think the federal government ought to be the primary agent for helping the poor.

Of course, some might take Aristotle's proverb to mean we should always punish bad behavior. No doubt, if the state fined or imprisoned men and women who had children out of wedlock or failed to get jobs when plenty were available, we'd have fewer of those problems. But few of us would want to live in a society so lacking in grace and compassion. Nor would most of us trust government to effectively distinguish the guilty from the victims, the sheep from the goats. So the question is, How do we help people in bad situations without encouraging bad behavior?

CIRCLES OF RESPONSIBILITY

The problem *isn't* that government workers are stupid or uncaring. The problem is all about information and incentives. A centralized government knows less about individual problems than does practically anyone closer to the problem. Replacing a family or a neighborhood or a local church with a federal program for helping the down-and-out is like trying to have an official in the Department of Commerce guess how much I should pay, right now, for a new pair of size-9 Asics running shoes. At the moment, I wouldn't pay much, since I just bought a pair. And I'm picky when it comes to colors. That, and I don't wear size 9! The official could look up the market price for Asics shoes in the United States. (It's between $60 and $140.) That's crucial information. He probably wouldn't know much else, though, so he'd just have to guess, and he'd probably guess wrong, and waste his time and mine in the process. That's the information gap in a nutshell. It's impossible to fix a problem if you don't know squat about it.

Think of responsibility as a cluster of overlapping circles or jurisdictions. The person or group with the narrowest jurisdiction has the most detailed knowledge and the most responsibility. The narrowest jurisdiction is the responsibility I have for myself. I know a lot more about what I need and when I need it than anybody else does. I know when I'm hungry, thirsty, and sleepy. I know when to breathe and blink. I know what I'm allergic to, what medicines I need to take and when. I know how to get to work, how to do my job, how to deposit paychecks in my account, how to recognize my friends, how to avoid that grumpy guy at the Northwest Airlines ticketing counter who doesn't like me. I know the names of my family, and on and on. I have privileged access to this information. Therefore, unless something bad happens, my responsibility for all these things should match the extent of my knowledge. I shouldn't expect someone else to handle all these details, since they don't know much about them.

If the mayor of Grand Rapids, Michigan, where I now live, suddenly had to assume responsibility for feeding me, giving me my medicine, and telling me when to breathe, he'd botch it, and I'd probably be dead, even if he's a member of Mensa and has a saintly disposition. And if he took over these responsibilities for years (assuming I lived), I might eventually forget how to do these things myself.

Of course, none of us is Robinson Crusoe; sometimes we need help. Parents take care of their children for years before those children can take care of themselves. Two-year-old Maria Gonzalez from Quana, Texas, isn't responsible for herself. Her parents are. That's the narrowest jurisdiction. It's only when parents are abusive, incompetent, or absent that other family members need to get involved. If no other family or neighbors are around, a church or nonprofit should step in. If that doesn't happen, the city or state government might have to step in. The federal government should be the *last resort*, because the more levels up the responsibility moves, the more Maria will have to be treated like a generic toddler: the more distant the jurisdiction, the less knowledge of the specific toddler. It doesn't matter how sweet employees at the Department of Health and Human Services are. They can't possibly know more about Maria's particular needs—her peanut allergy, her lack of sleep because of the barking Rottweiler next door—than can someone closer to the situation.

This is the problem with welfare dispensed by the federal government: it runs roughshod over this intricate web of overlapping responsibilities and assumes knowledge where none exists. When the federal government jumps in and ignores this intricate web, it violates the principle of subsidiarity. That's a fancy word for a simple idea. Here's how Pope Pius XI explained the principle:

> Just as it is wrong to withdraw from the individual and commit to a group what private initiative and effort can accomplish, so too it is an injustice for a larger and higher association to arrogate to itself functions which can be performed efficiently by smaller and lower associations. This is a fundamental principle. In its very nature the true aim of all social activity should be to help members of a social body, and never to destroy or absorb them.[24]

When the state takes over a task that is better handled by someone closer to the problem, it transgresses its proper boundaries and creates more problems than it set out to solve.

For the last forty years, we've reflexively looked to government to solve every problem, especially social problems. Child care, children's health insurance, rapid response to hurricane damage, homelessness, drug abuse—you name it, and somebody in Washington, D.C., is supposed to take care of it. Christian leaders have mostly gone along with this way of thinking.

HOW BIG IS TOO BIG?

Of course, there's nothing in the Bible that says, "Government expenditures must never exceed 15 percent of the gross domestic product. Twenty percent is an abomination to the Lord." The

Christian tradition doesn't give us any direct guidance on this, either. So how big should government be? That's a prudential question that doesn't have a unique answer. Still, it doesn't follow that there are no good or bad answers. Considering the massive growth of government in recent decades, the burden of proof should fall on those who call for even more government growth in areas outside its core competence.

Just consider the growth of federal regulations. According to a 2007 *Wall Street Journal* article, "It took 169 years from the founding for the federal code of laws to reach 11,472 pages— and only four decades more for that number to quadruple. In 1960, the Code of Federal Regulations numbered 22,000 pages; today that number has grown by more than 700%."[25] Moreover, government spending as a portion of GNP has grown exponentially in recent decades.

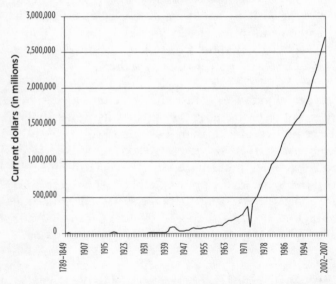

Federal Spending, 1789–2007

Figure 1. Office of Management and Budget, Executive Office of the President, *Summary of Receipts, Outlays, and Surpluses or Deficits: 1789–2007,* http://www.whitehouse.gov/omb/budget/fy2003/hist.html.

Government growth is even more staggering if you look at the federal budget. For much of its history, the federal government cost every citizen about twenty dollars a year (in current dollars, not the more valuable dollars of the past). Now it costs every one of us, on average, about ten thousand dollars.[26] And consider this chart of federal spending since 1789 (calculated in 2007 dollars). Look at the little spike after 1939, followed by a hockey-stick curve.

The pattern here is obvious: for a century and a half, government stayed about the same size; but in the last two-thirds of the twentieth century, it puffed up like a doughnut addict forced to work at Krispy Kreme. The story is much the same in the rest of the industrialized world. As the late Peter Drucker put it, by the 1960s "it had become accepted doctrine in all developed Western countries that government is the appropriate agent for all social problems and all social tasks."[27]

"If you want to encourage something, reward it. If you want to discourage it, punish it."
    —Aristotle

The problem isn't simply that taxes are too high. After all, not all forms of taxation are unjust. Every government has to collect taxes to fund services beneficial to all—to maintain courts, protect citizens from domestic and foreign predators, enforce traffic law and contracts, and so forth. These government functions stem from our inalienable rights. We have a right to protect ourselves from aggressors, for instance, so we can delegate that right to government. We *don't* have the right to take the property of one person and give it to another. Therefore, we can't rightfully delegate that function to the state. Delegated theft is still theft.

Using the state to *redistribute* wealth from one citizen to another is different from general taxation for legitimate governmental functions, such as those enumerated in the U.S. Constitution. Rather than promoting the general welfare, redistribution schemes involve a group of citizens voting to have the government take property from others and give it to them. Rather than celebrating such schemes, Christians should be holding them up to the light of moral scrutiny.

Besides being morally dubious, government transfers of wealth are degrading to recipients. As George Gilder put it in *Wealth and Poverty,* "It is extremely difficult to transfer value to people in a way that actually helps them. Excessive welfare hurts its recipients, demoralizing them or reducing them to an addictive dependency that can ruin their lives."[28] When government takes the property of one person and gives it to another, it sets up a lose-lose game disguised as a win-lose game. One group is coerced; the other is degraded.

Even if we forget these problems, however, government attempts to redistribute wealth simply don't solve poverty. This is now obvious—so obvious, in fact, that some enthusiasts for government welfare have begun to tame their rhetoric. Ron Sider and Jim Wallis, for instance, admit that government alone won't solve the problem. And they have talked a lot recently about partnerships between government and private charities. But they still haven't learned the basic economic lessons of the twentieth century, and they end up recommending the same old, same old. They still call for even more government involvement in everything from job training to wage controls to health care, all as part of a "new vision for overcoming poverty in America" (in Sider's words). Sider refers to this as "Just Generosity." But government spending on poverty isn't compassion or generosity. Compassion is a spiritual gift and implies that one "suffers alongside." And generosity is a free act. These are very different from government redistribution, because the money has been taken under threat of coercion, passed through a bureaucracy, and dispensed impersonally rather than given freely and directly.

Anticipating these points, some object that there's just not enough money in the private sector to take care of the poor and disadvantaged. As *Washington Post* columnist E. J. Dionne put it in *Sojourners:*

> Progressives . . . need to challenge a core conservative view that private and religious charity is sufficient to the task of

alleviating poverty. That is simply not true. In an important 1997 article in *Commentary* magazine—hardly a bastion of liberalism—William Bennett and John DiIulio made the crucial calculations: "If all of America's grant-making private foundations gave away all of their income and all of their assets, they could cover only a year's worth of current government expenditures on social welfare." What would happen the next year?[29]

Jim Wallis uses the same reasoning elsewhere in *Sojourners*.[30] Both conclude that however bad government welfare is, government *must* do these things, because the church and private sector simply don't have enough money.

OK. Let's pause and ask, *What are they forgetting?* First, they've assumed that all those government expenditures are actually helping people and ought to be spent. As we've seen, that's false. If anything, welfare has been a cultural wrecking ball in America's poorest communities. Second, Dionne and Wallis have posed a false dilemma—that the only possibilities are private charity *as it now exists* or government expenditures. But the money government spends comes from somewhere. It doesn't just appear magically in the federal treasury. The more money the government confiscates for counterproductive welfare programs, the less money is available in private hands—for families, investments, churches, and charities. The money can't be all places at once.

Third, they've ignored the fact that private charities *change* when government invades their territory. We often think of government welfare as a safety net to catch those missed by private charity. This way of thinking treats private charities as if they stay the same no matter what tasks the government takes on. That's false. At the Acton Institute, we give an annual prize of ten thousand dollars to the most effective private charity. To qualify, a charity can't receive more than 15 percent of its budget from government sources. In 2007, we had 311 applications from local organizations that shelter abused women, minister to runaways trapped in prostitution, care for foster children who fall between the cracks in the legal

system, help homeless people find homes and gainful employment, and so on.

Some private charities are more effective than others, but
almost all of them are more effective than larger government
programs. (If you're looking for a place to land your charitable
dollars, check out www.samaritanguide.org.) Still, all of these
organizations have one thing in common: they define their mission in part by what the government does or doesn't provide. If the
government weren't occupying most of the charitable ecosystem,
charities would be profoundly different. The ecosystem would be
filled with thousands of well-funded responsive charities accountable to their donors and communities. As it is, government has
invaded the ecosystem, and mostly made a mess of it.

Fourth, Dionne and Wallis ignore the difference between government programs and charity. Government programs produce
negative economic effects because of high taxes and perverse
incentives in its recipients. Private charities don't have the same
problems, since they receive money from voluntary philanthro-
pists keenly interested in results, and are therefore much more
accountable for producing results.

Finally, Dionne and Wallis assume that poverty will be al-
leviated by either charity or government handouts. Historically,
however, neither of these methods has done much to alleviate
widespread poverty. Private charity is better than big government when it comes to dealing with specific instances of poverty.
And it has proved itself extremely competent at providing disaster relief. But the only known cure for widespread, generational
poverty is capitalism. That's it. There's only one known path up
the mountain. Why do we think Africa or South Central Los
Angeles will get rich on foreign aid or charity when no society
anywhere has gotten rich this way? Why would we waste one
minute using those tools to fight poverty instead of working to
spread the one thing that we know works?

These are hard words. I'm *not* saying we shouldn't be charitable. Quite the contrary! We simply need to practice effective
charity while recognizing its limits. Charity, and sometimes
government action, are vital for extreme situations—such as

responding to humanitarian disasters in places like modern-day Sudan or in Western Europe after the defeat of Hitler. However, there's no evidence that either will solve the problem of poverty on a large scale and over the long term.

Charity is like gleaning in the Old Testament. When the Hebrews inherited the Promised Land, God commanded them not to reap the edges of their harvest, but to leave them, the "gleanings," for sojourners. The gleanings were tiny leftovers of a more bountiful harvest. They could never feed an entire city. That's a good model of charity: it's appropriate for emergencies and comes from a prior abundance. The more abundance, the more there is for charity. However, charity is limited. There's no evidence that entire cultures or nations, or even individuals, can or should grow prosperous by charity. Again, the only known way to create widespread wealth is through capitalism.

Of course, big campaigns to "end poverty as we know it" are hip. They inspire concerts and crusades. They warm our hearts. They give us a compassion fix. In contrast, the things that can actually help the poor sound about as glamorous as unsweetened oatmeal—property rights, rule of law, personal virtues like diligence and thrift, ingenuity, cultural values like trust, an orientation to the future, and a willingness to delay gratification. Yawn. Unlike the crusades and petitions, however, they create wealth. We know this. We Christians need to decide if we want to keep advocating what is hip and fashionable, or the oatmeal-variety stuff that actually works.

Choosing the oatmeal option behind door number two might seem to marginalize the Christian mission to help the poor, but it does not. Ministries that treat humans as fully spiritual beings rather than mere mouths to feed, that encourage economic freedom rather than government largesse, that teach the poor to fish, that instill Christian values—which ultimately transform culture—will do far more to reduce poverty in the long run than all the cool celebrity-led campaigns put together. Instead of being distracted by programs that don't work, Christian leaders need to serve as prophetic if unfashionable voices for the rule of law, justice, and human rights for all (none of which is a euphemism for government wealth-transfer schemes).

Admittedly, we lose something when we focus on capitalism rather than charity. When you give charitably, you feel good about yourself. That's not bad. But once we begin to understand that only the creation of wealth can alleviate poverty, and that capitalism is the best engine for creating wealth, we should be able to get some satisfaction from efforts to spread capitalism to places and people who are not yet benefiting from its bounty.

Three

Doesn't Capitalism Foster Unfair Competition?

After I became disenchanted with straight socialism, I still pined for a more just and humane society. Even if communism is bad, it doesn't follow that capitalism is good. And we all know that capitalism means dog-eat-dog competition, survival of the fittest, and all that, right?

That's certainly the general impression. A century ago, Yale social scientist William Graham Sumner defended capitalism as a model of Darwinian natural selection. As he explained it, capitalism is all about weeding out the weak and favoring the strong: "Millionaires are a product of natural selection," he explained. If we find survival of the fittest distasteful, "we have only one possible alternative, and that is the survival of the unfittest."[1] He criticized welfare programs as an "absurd attempt to make the world over."

American novelist Jack London described *"laissez faire capitalism"* as "everybody for himself and devil take the hindmost. . . . It is the let-alone policy, the struggle for existence, which strengthens the strong, destroys the weak, and makes a finer and more capable breed of men."[2]

In the same vein, former Democratic presidential candidate Walter Mondale once criticized Ronald Reagan's economic policies by saying he believed in "social decency, not Social Darwinism."[3] And in his encyclical *Quadragesimo Anno,* Pope Pius XI said:

Free competition, though justified and right within limits, cannot be an adequate controlling principle in economic affairs. This has been abundantly proved by the consequences

that have followed from the free rein given to these dangerous individualistic ideals.[4]

These aren't baseless complaints. When you first think about
it, it seems obvious: If I'm competing for business and I succeed, someone else loses. Therefore, if I make money on a business venture, someone else must have lost money. My success,
then, is tied to someone else's failure. Gordon Gekko, the ruthless high-stakes investor in the 1987 movie *Wall Street,* said it
better than anyone: "It's not a question of enough. . . . It's a
zero-sum game. Somebody wins. Somebody loses." From there,
it's a short step from individuals to nations: strong nations like
the United States win; weaker countries like Nicaragua lose
and lose and lose.

If that's right, then a market that allows some to get richer
than others is basically theft on a large scale. The strong prey on
the weak; end of story. The details might be hazy, but the basic
idea is clear enough. At least it seemed clear to me. I found it
hard to shake this view of capitalism, but I slowly learned that
the obvious is not always true. And in the process, I also learned
that lurking among the turgid truths of economics is one of the
greatest mysteries in the universe.

"Economics is the study of mankind in the ordinary business of life."
—Alfred Marshall

EVERYTHING I NEEDED TO KNOW ABOUT THE MARKET I LEARNED IN KINDERGARTEN

OK, not everything. And it wasn't kindergarten. It was sixth
grade. We played the "trading game." I didn't know it had a
point. I figured our teacher was just distracting us on a cold
snowy day, since we couldn't go outside during recess to play
something really useful, like kickball. The teacher passed out

little gifts to all of the students: a ten-pack of Doublemint gum, a paddle board with one of those little rubber balls tied to it, a Bugs Bunny picture frame, an egg of Silly Putty, a set of Barbie trading cards—stuff like that. The teacher probably picked them all up in the dollar aisle at the local Kmart.

After giving every kid one gift, our teacher split the class into five groups of five students each. She then asked us to write down, on a scale of one to ten, how much we liked our gift. That was it. We didn't have to talk to anyone else. We just decided how much we liked our toy. Our teacher then compiled all the scores and added up the total.

Then she said we could trade with the other kids in our group. No one had to trade, but if I had Barbie trading cards (which I did), and Sarah had a paddle board (which she did), and each of us preferred the other's gift (which we did), we were free to exchange. Of course, a few students kept or got stuck with their original gift, but many ended up with a toy they liked more. Again, we graded our toys and the teacher added up the scores. The total score went up.

Then she told us that we could trade with everyone in the room. Now we all had twenty-four possible trading partners rather than just four. Almost everyone, including kids who a minute before were as lively as a marshmallow, was suddenly cobbling together complex trades that would have made a game theorist proud.

After everyone had a chance to trade, we again graded our toys and added up the scores. The total number had gone way up. Almost everyone ended up with a toy he liked more than the one he started with. No one had a score that had gone down. The only kids whose scores didn't go up were the ones who happened to get gifts they really liked at the beginning. I got the paddle board on the first trade and kept it.

I didn't get the point of the game until I played it again twenty-five years later. As it happens, the game teaches some of the most important lessons of economics.

Lesson One: Trading freely can add value even though the traded items remain physically unchanged.

In the game, notice that nothing new was added after the teacher handed out the gifts. They just got reshuffled freely by the players. And yet the total value went up. By itself, that's interesting. To find out how it happens, see Lesson Seven.

Lesson Two: Normally when trading freely, the more trading partners there are, the better.

Notice that in the trading game, the total value of the gifts—that is, how much everyone liked the gifts they had—*went up* with the number of possible trading partners. Some of the students graded their final toy as a nine, while others graded theirs as a six. So everyone didn't end up equal at the end. Still, almost everyone ended up with a gift they liked more than the gift they started with, even though nothing new was added to the game and the players were competing with each other.

We experience something like this every day with the Internet, which illustrates Metcalfe's Law. Metcalfe's Law (named after electrical engineer Robert Metcalfe, who first suggested it) states that the value of a telecommunications network increases with the square of the number of users. The basic idea is that, because of all the multiplying interconnections as you add users to a network one after another, the value of the network goes up not linearly but exponentially—that is, a lot rather than just a little bit. The same is true in the trading game. The point is obvious when the game is relatively small and you can see all the players. But somehow we miss the point when the game goes global.

Lesson Three: A free exchange is a win-win game.

There are three kinds of games: win-lose, lose-lose, and win-win. Win-lose games, like chess, checkers, poker, basketball, and badminton, are sometimes called *zero-sum games*. When the Celtics and the Bulls compete, if the Celtics are up, then the Bulls are down, and vice versa. The scales balance. It's a zero sum.

Besides win-lose games, there are the never popular lose-lose games, like all-out nuclear war in a confined space. You don't see these games played very often, because no one wants to play them. In the rare cases where a lose-lose game is played, it's hard to get participants to describe the experience, either because they're humiliated or because they're dead.

Then there are positive-sum or *win-win* games. Nobody loses. In win-win games, some players may end up better off than others, but everyone ends up better off than they were at the beginning.

An exchange that is free on both sides, in which no one is forced or tricked into participating, is a win-win game. It's a positive-sum game. Think about it. The trading partners wouldn't trade if each did not perceive himself as better off as a result. In the trading game, I preferred the cool paddle board. Sarah preferred the stupid Barbie trading cards. Even though nothing new was added to the system, the outcome was still a win-win. Understand this one simple fact, and you understand economics better than most critics of capitalism.

Lesson Four: The game is win-win because of the rules set up beforehand.

The players aren't allowed to coerce or steal from each other. A free market is *not* a free-for-all in which everybody can do what they want. That's anarchy, in which the strong can steal from the weak. That's raw competition, survival of the fittest, nature red in tooth and claw. A free market isn't like that. Any exchange must be free on both sides. The participants are free to exchange or not to exchange. Ideally, the players would be virtuous enough to play by the rules. But in the trading game, as in real life, you can't count on the virtue of others. There's a bully in every class. So you need an outside enforcer. In the game where I was able to unload the Barbie trading cards, the teacher played this role.

In real life, it includes parents, teachers, churches, that old lady down the street with the steely gaze, and the government.

Lesson Five: Scarcity is almost always real.

If there was a big bucket in the room filled with all the same toys, no one would bother to trade. But in most of life, there is almost always scarcity. The trading game is like the real world of trade in this respect. And where there's scarcity, there's competition. We can't do anything about that. It's how the world is. The peaceable kingdom of God with limitless plenty isn't one of the options. The basic options are a win-lose society based on the laws of the jungle, a lose-lose society of coercive socialism (more on this later), or a market where win-wins are possible. The right rules can make the trading game win-win, even when there's scarcity.

Lesson Six: Opportunity costs.

When there's scarcity, there's always a trade-off. In the trading game, no one gets to have more than one toy. In the real world, if I have only ten grand, I can't buy both a ten-thousand-dollar Chevy Impala and a ten-thousand-dollar first-issue *Tales from the Crypt* comic book. I have to choose one or the other. A trade-off is, in a sense, what something costs. Trading can make me better off than I was before, but it can't give me everything.

Lesson Seven: Economic value is in the eye of the beholder.

For centuries, economists and philosophers were confused about economic value. This isn't surprising. It took us centuries to unlock just a few of the mysteries of physics and biology. Economics is much more complicated, since it deals with the complex interactions of the most complex things in the physical universe—human beings. So thinkers as different as Adam Smith and Karl Marx defined value in terms of labor or cost of production.[5] They thought that how much it costs to produce something determines how much it is worth. This meant that with enough information you could calculate the true or "just" price of just about anything.

Some still believe this. For instance, a recent article in *Sojourners* magazine said: "One of Karl Marx's more reasonable ideas held that the value of a commodity was comprised of the labor that went into it."[6] Well, it's "reasonable" in the sense of "wrong." Even Marx's loyal followers quickly dispensed with this theory, but the *Sojourners* writer apparently didn't get the memo.

Think of a real-estate developer who buys a plot of land and plans to build ten new houses on it, each a different color. He then hires a bunch of construction workers and pays them all twenty dollars an hour. One of the houses, the pink house, takes ten laborers twelve months to build, while another house, the yellow house, gets put together by three men in a month. So the pink house cost much more to build than the yellow house. For the developer to meet his costs, then, he'll have to charge far more for the pink house than for the yellow house.

Let's say he adds up all the salaries of the laborers who built the house, and then treats the price of raw materials and necessary tools as the "labor" of other workers. He even includes his own salary. Let's say it cost him five hundred thousand dollars to produce the pink house, but only two hundred thousand dollars to build the yellow house. So the pink house will be worth more than the yellow house, right?

Wrong. The fact that it cost more to build the pink house doesn't mean that anyone will prefer the pink house over the yellow house. Maybe the expensive pink house was built on a rocky lot next to an ugly landfill and lavished with those dingy brushed-bronze fixtures last stylish in the late 1970s. Or maybe nobody in town wants a pink house. Or maybe it stinks inside. Or maybe the pink house just costs more than anyone is willing to pay. In any case, the developer can slap "$500,000" on the For Sale sign, but that doesn't mean anyone will pay up. What sense does it make to say the house is "worth" five hundred thousand dollars if no one will pay that much for it?

In truth, the developer failed to gauge the market properly. He didn't notice that he had put his ten slowest workers on the same project. He made a mistake by building a house that no one wants, at least not for five hundred thousand dollars. His cost exceeded the economic value of the house. How does he find out what the house is worth? Well, following the trusty rule of supply and demand, he can keep lowering the price until he finds a buyer. If he eventually finds someone who will pay two hundred thousand dollars for the house, he's finally discovered what it's worth. That's because the economic value of something is determined by how much someone is willing to give up to get it.

Of course, labor often adds value to a product, as long as that labor creates what someone wants. But you can't *define* economic value in terms of labor. Someone can dig a ditch in a field until his hands are bloody and raw, and then fill it back in again, without making anything that anyone wants. In that case, there's a lot of labor, but no economic value. The refilled hole is still worthless, even if somebody pays the digger a hundred dollars an hour to dig and fill it.

This may have been Marx's biggest blunder. I say this because Marx's prophecy that capitalism would destroy itself hinged on his labor theory of value.[7] According to Marx, when a factory owner hires a worker to build a chair and then sells the chair for more than it cost to produce, the owner has taken more than the good is actually worth. He's taken its "surplus value." Such profit is basically theft, since, on Marx's terms, the chair is worth exactly what it cost to produce it. So the factory owner has gotten more than it's really worth. This is why Marx speaks of capitalists "exploiting" workers, even if the workers have chosen to work for the salary they are given.

Without his definition of value, however, Marx's argument collapses. The workers have received what they agreed to. The factory owner has wisely combined their labor with his resources. He then markets and sells the chairs for more than they cost to produce but not more than others will freely pay. He's rewarded with profit for his entrepreneurial effort. There's no injustice here, no exploitation, no contradiction that will inevitably lead to class warfare and revolution.

This may sound like philosophical nitpicking, but millions of people died in the twentieth century, in part because of Marx's error. And it's not yesterday's news: millions of Christians still don't get economics for the same reason. They compare the high salary of a business owner (who bears the risk) or a CEO (who may make decisions worth many millions of dollars) with the lower salary of an employee and assume that somebody's getting the shaft. "It's a 'fraud,'" said Jim Wallis recently, "when the average CEO of a Standard & Poor's 500 company made $13.5 million in total compensation in 2005, while a minimum wage worker made $10,700."[8] That's an assertion, not an argument. Unless those workers were lied to or forced to work, or the CEO lied about his record, there's no fraud here. Everyone has freely chosen to work for their wages. Of course, some of those workers may have limited choices. That's regrettable, but it's not fraud.

Economic value is not only in the eye of the beholder; the beholder's eye can see things differently in different circumstances. I'll value the same glass of water much more if I'm hot and thirsty, and miles from any other water, than if I'm sitting in my air-conditioned kitchen, where I can get a glass of ice water anytime I want. And I'll value the tenth refill of that glass less than the first, even though the same amount of labor was required to make each of the glasses available to me.

Even though I need water to survive, I usually can get water whenever I want it. So it's practically free. In contrast, a painting by Van Gogh, which no one really needs, is still really expensive. Part of the reason is because the paintings are scarce. Unlike with water, you can't get a Van Gogh free from a Dutch-impressionist dispenser in a department store. But that's not the whole story. Van Goghs aren't only scarce; they're desired. Lots of people will cough up serious dough to get one of them. And this affects their price for everyone. That's why I'd almost always prefer a Van Gogh to a glass of water. Only if I were dying of thirst would I prefer a glass of water to *Starry Night,* since in other circumstances I could exchange the painting for millions of gallons of water anytime I wanted.

Adam Smith, who held an on-again-off-again labor theory of value, struggled to resolve this paradox in his *Wealth of Nations.* It seems like a paradox if you're looking for economic value in the wrong place. But it's not a paradox once you realize that economic value has to do with how much I, and lots of other people, value goods and services in the marketplace.

These insights were first developed by Christian scholars in the middle ages,[9] but most Christian critics of capitalism still don't know about them. Others have heard them but dislike them. Part of the blame for that lies with economists themselves, who often talk as if economics explains everything. For instance: "The measure of value," said Austrian economist Carl Menger, "is entirely subjective in nature."[10] This sounds a lot like Protagoras's claim that "man is the measure of all things."

But the word *value* is fuzzy. It has several meanings. To say *economic* value is subjective is not to say, with many a college freshman, that "everything is relative." We're talking about *economic* value, not ultimate value. My ultimate value in the eyes of God is not the same as my *economic* value. Our true value is not found in our intelligence, skin color, good grooming, strength, or stock-market prowess. No one has the authority to make one person more valuable than another. "All men are created equal"—equal in value and dignity—because all of us are created in God's image. Our equality doesn't vary with the wants and needs of consumers in the marketplace. So ultimately, the small handicapped child in rural India is worth just as much as Bill Gates, who has created millions of jobs. Gates's wife, Melinda Gates, gets this: "One life on this planet is no more valuable than the next."

But Bill Gates and that handicapped child don't have the same economic value. In fact, in economic terms most children are liabilities rather than assets. Economist Gary Becker has compared children to houses, because they are "expensive to produce and maintain" and to refrigerators, because "they have a poor secondhand market."[11] But only a nut would conclude that his young daughter was therefore worthless. There's more to life and reality than economics.

Myth no. 3: The Zero-Sum Game Myth (believing that trade requires a winner and a loser)

Free trade makes it possible for people to play win-win games of exchange. We're so used to this that we take it for granted. But imagine the alternative. Let's go back to the stinky pink house. Say the pink house costs five hundred thousand dollars, because that's how much it cost to build it. And let's say the labor theory of value is the law of the land. Then, for the developer to receive a "just price" for the house, he must receive five hundred thousand dollars. But what if no one wants to pay that much for the house? Then the only way for him to get his "just price" is for someone to be forced to pay that price, whether the person likes it or not. That's not a win-win. That's a win for the developer, and a big fat loss for the new, reluctant homeowner. And eventually it would be a loss for the entire economy. Here's why: In a society that worked this way, developers could build whatever they want, spend as much as they want, ignore what other people want, hire lazy and lying workers, and still get fully compensated. A forced market based on the labor theory of value would lead directly to a world of crummy housing. If this sounds like an academic fantasy, go check out the housing developments built under the Soviet Union. They make the stinky pink house look good.

Let's dispense with a forced housing market based on a bogus theory of value and free things up a bit. If the developer can't make you buy his house, he's going to be more careful to build houses that people will freely buy. He's going to look for ways to set up a win-win situation. He's going to think about what other people want rather than what he wants to build. He's going to look for ways to build houses below the price people are freely willing to pay. Even if he's selfish, he's going to be directed toward meeting the needs of others. If he does this well, he may get fabulously rich, but none of his customers should begrudge him his wealth, because they all got something that they wanted more than the money they gave him. Win-win.

Ditto when it comes to employment. If you aren't forced to take a job building a Web site for an ad agency, you'll take the job only if you think you'll be better off as a result. And the agency won't hire you unless they believe they'll be better off. So it's a win on both sides. You shouldn't compare who's better off as a result of the deal, you or the ad agency. That way lies envy. Since it's a free exchange, you should compare your status with the job versus your status without it.

Of course, this doesn't mean that in a free market no one ever loses. After all, what about the other guys competing for the same business? Go back to the example of developers. There are only so many customers looking for houses. And if our developer sells his houses to ten people, there are ten fewer customers for everyone else trying to sell houses. Surely it's a big fat loss for them, isn't it? Obviously they will lose those sales. A free market doesn't guarantee that everyone wins in every competition. Rather, it allows many more win-win encounters than any alternative.

That said, remember that the other developers are also potential customers. Even they may benefit directly if they buy the better, cheaper houses sold by their competitor. Or they may benefit indirectly as homebuyers because the market is competitive, so houses will cost less than they would otherwise. So in a market, competition is almost always better than monopoly, since competitors generally will focus on meeting the wants and needs of consumers rather than on stealing from them.

FROM TOYS TO LIFE

Of course, a real market is much more complicated than twenty-five students swapping toys from the dollar bin. In the trading game, players can trade only toys. It's a primitive barter system. This kind of bartering works only if you have what someone else wants, and in the right amount. For instance, let's say you have one cow and your neighbor has a pineapple farm. You want to buy a pineapple from him. The two of you might agree that a cow is worth a hundred pineapples. But what do you do if you don't want a hundred pineapples and your neighbor won't take one one-hundredth of your cow? You're stuck.

In real life, we have money that can stand in for our assets. Money is like trade lubrication. It solves the cow/pineapple problem because it's a standard medium of exchange that can be multiplied or divided into different-sized increments. This makes trade much easier and creates far more choices than students have in the trading game.

Also, in real life we exchange not just our stuff but our labor, our skills, our ideas, and our creations (more on that later). Within limits, it's best for trading partners to specialize, to focus on their "competitive advantage." Your competitive advantage might be based on skill, luck, or location. Learning to fly a Boeing 747 gives you a competitive advantage as an airline pilot over most other people. That's a skill. If you're seven-foot-nine and even slightly coordinated, you've got a competitive advantage over me in playing basketball. That's luck. And if you're Guatemala, you've got a competitive advantage over Norway when it comes to growing bananas. That's location. Add the sheer size of a global market, and the possibilities for enrichment through free trade are immense.

Despite these differences, however, the same rules apply to both the trading game and a free market. In the trading game, players start out with toys they can trade. For a market to work right, people need private property. That includes not just our stuff, but our labor, our time, and our ideas. So a society that respects private property must have laws against theft, fraud, kidnapping, and slavery. (Slavery is, among other evils, stolen labor.)

In some form, private property has been around since some beleaguered Cro-Magnon staked out a dry cave to store his choicest mammoth steaks. Still, formal property law is most developed in modern Western cultures. That's no surprise, since it's well grounded in the Bible, which has deeply influenced the West.[12] In the Bible, the right to private property is nowhere stated but everywhere assumed. The Ten Commandments—a sort of summary of all God's laws—take private property for granted. For instance, the eighth commandment, the one against stealing, implies that we may have property. Otherwise, there would be nothing to steal, and the commandment would make no more sense than an order not to fraternize with four-headed Jube Jube monsters. (No, I don't know what they are, either. I just know they don't exist.) The tenth commandment, against coveting, also assumes that we can have stuff. After all, you can't covet your neighbor's oxen if your neighbor doesn't own oxen.

The roots of private property reach all the way back to Genesis. The author of Genesis got the mother of all writing assignments. In just fifty short chapters, he had to describe everything from the creation of the universe to the Jews' arrival in Egypt with Joseph. With such space limits, some periods are summarized. Everything from the beginning of the universe to the creation of human beings, for example, gets twenty-five verses. And yet one obscure incident involving the purchase of land gets a whole chapter.

In Genesis 23, after Sarah dies, Abraham asks the Hittites for a plot of land to bury his wife. The Hittites are gracious and offer him any spot he wishes. Abraham surveys the land and chooses the "cave of Machpelah, which [Ephron the Hittite] owns . . . at the end of his field." Ephron offers to give it to Abraham, but Abraham insists on buying the plot for the full price—four hundred shekels of silver—at the city gate in plain view of the other Hittites. And then we read: "So the field of Ephron in Machpelah, which was to the east of Mamre, the field with the cave that was in it and all the trees that were in the field, throughout the whole area, passed to Abraham as a possession in the presence of the Hittites, in the presence of all who went in at the gate of his city" (Gen. 23:17–18). Why all the details?

Genesis is a grand epic that starts with God creating the universe and quickly narrows to the history of the Jews. But in chapter 23, it's like you've stumbled into some Bronze Age real-estate deal. One respected biblical scholar called the passage "pedantic and almost comic."[13] Not the stuff of epics for sure. But the author of Genesis thought these details were worth recording. And they were. Here we see Abraham acquiring permanent legal title to the land. This may seem trivial—until you compare societies that have formal rules for titling with those that don't. If you live in the United States, Europe, Japan, or a former British colony, you probably take such procedures for granted. In the United States, every square inch of dirt from San Diego, California, to Caribou, Maine, belongs to somebody. And it's easy to verify. But many parts of the world today still lack rules for buying and titling land as formal as those recorded in Genesis 23. Most people in those parts of the world are dirt-poor for the same reason.

Peruvian economist Hernando de Soto has argued that a formal property system is the key that unlocks the mystery of capital. And he says that this helps explain why capitalism has succeeded in the West but not in many other places.[14] In much of the developing world, people may cultivate and live on land that they don't clearly own, or own only "extralegally." They don't have titles representing their ownership, documents that the legal authorities will protect and accept as binding. As a result, a farmer may toil over the same plot of land for decades but never get beyond hand-to-mouth subsistence farming, producing just enough to survive. The land serves only its immediate physical purposes, and doesn't become a real asset. The farmer can't exploit its untold potential. He acts for the short term, not the long term. Since it is held informally or "extralegally," de Soto calls such land "dead capital."

His research team has calculated that "the total value of real estate held but not legally owned by the poor of the Third World and former communist nations is at least $9.3 trillion."[15] That's vastly more than all the foreign aid ever given to these countries, much to little effect. Third-world people may "have houses but not titles, crops but not deeds, businesses but not statutes of incorporation."[16]

To add insult to injury, these countries have bureaucratic roadblocks that look more like torture devices. They seem designed to keep citizens from opening businesses or owning property legally. To buy a home legally in Peru, for instance, you'll wander through a five-stage maze, with 207 steps in the first stage alone.[17] In Haiti, you have to jump through 65 bureaucratic hoops just to lease land from the government. Buying the land? Another 111 hoops. It would take a diligent Haitian nineteen years to pull this off, assuming he could get the time off from work, if he has work. And the fact that he's acquired property legally doesn't necessarily mean that his situation won't change tomorrow, since the country itself frequently changes owners.[18]

DeSoto then compares this to the situation in developed Western countries. Here, "the same assets also lead a parallel life as capital outside the physical world. They can be used to put in motion more production by securing the interests of other parties as 'collateral' for a mortgage, for example, or by assuring the supply of other forms of credit and public utilities."[19]

When I buy a house in Grand Rapids, Michigan, nothing about it has changed physically. It hasn't gone anywhere. We haven't anointed it with oil or performed a pagan ceremony over it. We've just represented it in a new way. We've recorded and fixed its economic value on paper or computer and embedded that information into a much larger system of laws and agreements. This creates strong bonds of accountability enforced by law. I'm accountable to Fifth Third Bank, which gave me the loan. My credit rating depends on my paying my bills on time. My neighbors are accountable to me. However rich or strong they are, they can't legally steal my house. I'm accountable to them, since I can't do anything with my property that harms theirs. Even the government is restricted. If it wants to build a highway through my front yard, it will have to compensate me.

As a result, new possibilities emerge. I can now get utilities like water and electricity. (You think scheduling the cable guy is hard. Try getting utilities without a legal address.) And without too much trouble, I can use the equity on my house to get a loan to partner with somebody to open a new Starbucks, since my block doesn't have one—yet. Without that collateral, the bank would probably not take on such risk. My house has been *transformed* into more than just a house. It's an asset. It's capital. All this simply because we represent it in a certain way.[20] Magnificent.

Of course, formal private property isn't a magic bullet for all the world's ills. Economically, it's just the beginning, the first rung on the ladder. Without it, an economy will rarely be win-win for its participants. Once you've got it, though, weird stuff starts to happen.

THE MYSTERY OF THE MARKET

In his *Wealth of Nations,* Adam Smith described the market as an "invisible hand" that guides the acts of individuals "to promote an end which is no part of [their] intention." Often that end is better than if the individual members of a market had tried to bring it about.[21] The end Smith had in mind is an orderly and efficient distribution of goods and services that emerges from the generally self-interested acts of people producing, buying, and selling in a market without fixed prices, even though no one intends such an order.

Smith believed in God, so he saw this invisible hand as God's providence over human affairs, since it creates a more harmonious order than any human being could contrive. Although Austrian economist F. A. Hayek did not see God's providence in the market, he, too, marveled at what he called its "spontaneous order."[22]

Many think that a free market will lead to chaos if left to its own devices, and that justice and order require that someone, usually the government, keep it in check. In the 1980s, the final premier of the Soviet Union, Mikhail Gorbachev, traveled to Great Britain and met British Prime Minister Margaret Thatcher. While touring with Thatcher, a perplexed Gorbachev asked Thatcher who saw to it that the British people got fed. "No one," she told him. "The price system does that." Untutored common sense sides with Gorbachev; reality, with Thatcher. The British people get fed, and far better than under the hypermanaged economy of the Soviet Union. So obviously common sense needs tutoring.

In a famous conversation between Mikhail Gorbachev, the last premier of the Soviet Union, and Margaret Thatcher, then British prime minister, a perplexed Gorbachev purportedly asked Thatcher who saw to it that the British people got fed. "No one," she told him. "The price system does that."

At the same time, the hunch that some sort of oversight is required to make a market work is on target. Free trade isn't anarchy. It requires a rule of law that makes sure one person doesn't steal from another person or force the other person into an unwilling exchange (merely a more sophisticated form of theft). But that doesn't mean the government should control the market. Think of the trading game again. What if, instead of the students getting to trade freely, the teacher dictated or tried to guess which toy each of her twenty-five students preferred? What are the chances that the students would end up with toys they liked just as much as the toys they got by trading freely? Near zero. What if the teacher was a substitute teacher who had never met the class before? Nearer zero. Now multiply the problem by a few bazillion, and you have some sense of the problem confronting anyone who wants to centrally plan an economy, as the Socialists did.

To plan a whole economy, you have to set prices and production quotas for all the goods and services it contains. Do this even a little wrong, and there will be all sorts of wasteful surpluses and shortages in the market. But nobody has ever succeeded in planning a whole economy of any size and gotten it just a little wrong. Either they quickly backed away from the attempt to plan the entire economy, or else they created widespread famine and death, as Lenin and Mao Tse-tung discovered. Why did they fail? It's not because these men were stupid. It's because they didn't know it all. That's pretty much what it would it take. The successful master controller would need to know the economic value not just of every product in the market. He would need to know the economic value of every individual thing in the market at any given time and place, since the value of things can change drastically depending on the situation. Remember, the value of a good is equal to the wealth and opportunity someone is willing to give up to obtain it at a specific moment. Such assessments vary from person to person and vary even for the same person in different circumstances and at different times. No one has access to all of that information.[23]

The issue is not just the sheer number of choices. If this were only a math problem, someday a bigger, faster computer might solve it. The problem is knowledge. No one has access to all the constantly changing judgments of the billions of agents involved even in small economies. The conclusion is inescapable: central economic planning is impossible—full stop.

But if no one can know the economic value of anything before the moment of free exchange, how is trade even possible? This is the central mystery and wonder of the market order: we *can* learn the economic value of everything—from a February car wash in Grand Rapids, Michigan, to a pair of men's size-11 Nike Air running shoes—but *only* through the market process that leads to a specific price.

This isn't because humans are fallen and often have warped ideas about the worth of things. Even if the entire world were filled with entirely sanctified Christians, the market would still be the only way to get the needed information.[24] If a Chick-Fil-A sandwich is subject to the market, its price tells me what I need to know about how scarce such sandwiches are and how much other people value them. If PETA decides to target Chick-Fil-A sandwiches for propagating violence against chickens so that fewer animal lovers want the sandwiches, that fact will be reflected by a drop in the price. If a chicken plague wipes out half the chickens on the planet and everything else stays the same, the price for Chick-Fil-A sandwiches will go way up.

The price of a good or service in a free economy, then, is a little packet of information that tells us its economic value at that moment. It represents an underlying reality, not merely the random choice of a store clerk. So a market doesn't just distribute goods and services. It's a highly sensitive network for gathering and disseminating information that would otherwise elude us. It leads to specific prices for the goods and services of interest.

The price alone allows entrepreneurs to decide where to invest, and it tells producers how many chicken sandwiches to produce compared with other things it could produce.

In a free market, goods and services generally end up where they're most valued. If a hurricane in the South leads to a local shortage of gasoline, gas prices will go up there, drawing gasoline where it's most wanted and away from Bismarck and Boston. As gas gets scarcer in those places, the prices will go up there. That price increase will draw some of the gas back, so all the gas won't flee to Biloxi and Baton Rouge. In short, what emerges from a free market regulated by prices is not utopia, but it is an order more complex and efficient than any order that human planners have or even can devise.

I, IPOD

In 1958, a man named Leonard Read wrote a little essay called "I, Pencil."[25] He wrote it from the viewpoint of "an ordinary wooden pencil" that was explaining its origins. We think a pencil is simple. And yet a single pencil requires what Read calls "innumerable antecedents," involving millions of people, from Albania to Zimbabwe, performing all sorts of different tasks. First, there's the cedar tree harvested from northern California. Then there are the saws and trucks and ropes and other equipment, all built in different places; the mill in San Leandro, California; the trains to transport the wood; the processing plant with kiln and tinting; the electricity from the dam to power the plant; the millions of dollars in equipment used to build the pencils; the graphite from Sri Lanka, mixed with clay from Mississippi and chemicals from who knows where; the wax from Mexico and beyond; the yellow lacquer with castor oil; the brass to hold the eraser, forged with metals from mines from around the world; the eraser made with factice from Indonesia and pumice from Italy. Finally, there are the trucks that deliver the pencils and the stores that sell pencils to the public for about ten cents each. All of this and much more is needed to make one yellow pencil.

There are at least three marvels here. First, few who contributed to the pencil meant to a make a pencil. The miner in Sri Lanka probably doesn't know that the graphite he's mining will end up as a pencil, and the miners who mined the copper to make brass probably don't know that the little metal thingy on pencils is even made of brass.

Second, as the pencil in Read's tale explains, *not a single person on the face of this earth knows how to make me.*" That knowledge isn't stored in any one place but is dispersed among millions of different people.

Third, no (human) mastermind oversees the process. It's all coordinated by people working freely in specialized jobs following the price of countless goods and services. This is one of the greatest wonders in the universe, but it's gotten the worst press. If we didn't see its effects after the fact, we would never believe it. Even seeing isn't believing for some people. Science writer and socialist Isaac Asimov scoffed at the market: "I cannot understand it," he admitted, "and I cannot believe that anyone else understands it, either. People may say they understand it . . . but I think it is all a fake."[26]

Most of us believe pencils really exist, though, even if we don't know how they're made. Imagine if Leonard Read had written the essay today and chosen an Apple iPod rather than a pencil. I have an iPod nano. I love it. I haven't the slightest idea what makes the darn thing go, let alone how to make one from scratch. No surprise there. But the fact is, *no one* does.

An iPod is Apple's nifty MP3 player. An MP3 is a compressed audio file that can be transmitted over the Internet and stored on computers. Think of it as a little digital "recording" of a song or a lecture. Beyond that, it's hard to describe in plain English. Even the physical parts of the iPod are hard to describe. The layman is reduced to naming the companies involved. The regular iPod has a hard drive with spinning disks. My nano, which is about the size of five credit cards pressed together, uses something called flash memory, developed by Samsung in South Korea, which stores information in floating-gate transistors. It can hold billions of bytes—little units of information.

The iPod's sophisticated user interface, developed by a company called Pixo, is an ingenious touch pad shaped like a doughnut. You can find any file on your iPod simply by rubbing your finger on the circle and pushing it at different spots. You view the menus on a color liquid crystal display with an LED backlight. Dozens of engineers and designers contributed to this simple navigation tool.

Its central processing unit—the chip that serves as the brain for the machine—is made by Samsung. The audio chip is made by Wolfson Microelectronics. The whole thing runs on a three-millimeter-thick rechargeable lithium polymer battery made by Sony. It's covered in a stainless-steel and plastic casing, which rarely breaks when dropped. Little earbuds plug into the iPod and deliver sound to your ears. The iPod connects to computers, which are connected to the World Wide Web—a network of computers, satellites, fiber-optic cables, and other cool stuff that blankets the planet. In this way, my iPod communicates with a free Apple program and network called iTunes. iTunes lets me search for audio and video files and buy them with the click of a button, usually for 99 cents. (Boring lectures are usually free.) Tens of thousands of patents lie behind the technology. Millions of engineers, technicians, and other workers around the world work to make all this possible. None of them knows how to make an iPod from scratch.

It would take pages and pages for me to describe just a tiny part of the iPod. If we traced the antecedents beyond the computer and software companies, however, to the delivery companies, mines, patent offices, and power plants, you would find hundreds of millions of people working in seemingly unrelated jobs, speaking dozens of languages, all blissfully unaware that they are, in a small way, making the iPod possible. And yet no one is or even could oversee this whole process, not even Apple Inc. As one commentator put it: Apple "may not make the iPod, but they created it. In the end, that's what really matters."[27]

That's the free market. We hear a lot about the brutish, competitive nature of capitalism, about winners and losers, survival of the fittest, and all that. Some of us may even have downloaded a podcast on the subject right onto our iPods. We hear far too little about the miracles of free cooperation and interdependence that free markets have made possible, that have helped make things like the iPod possible. Whatever the other vices in the market, we should take no critic seriously who does not first recognize this virtue.[28]

> "I cannot understand it, and I cannot believe that anyone else understands it, either. People may say they understand it . . . but I think it is all a fake."
> —Socialist writer Isaac Asimov, expressing his surprise at the effects of the free market

LONG WAY TO GO

Healthy capitalism isn't governed by the law of the jungle, where the strong destroy the weak. As long as there's scarcity, there will always be competition, but in a market economy marked by the rule of law, stealing, fraud, slavery, and kidnapping won't be the typical way to get ahead. Instead, most people will try to meet the needs and wants of consumers. And everyone is a consumer. The logic of competition in a market is not about destroying enemies. It's about serving consumers better than your competitors. Where prices can roam freely, goods, services, and information will generally be well distributed. In fact, for this purpose the market has no known competitors.

Finally, if trade is free, people will look for ways to engage in win-win exchanges. This means that if you're rich and someone else is poor, you could have grown rich by meeting the needs of others rather than by impoverishing them. You might benefit much more than your trading partners, but even they will be better off as a result of the trade. So capitalism doesn't foster *unfair* competition.

But what if your trading partner makes only two dollars a day? True, she might make only one dollar a day without free trade, but that's not much consolation, since she's still dirt-poor.

So far, we've answered the letter but not the spirit of the question that opens this chapter. There's a deeper issue. It's the link we see, or think we see, between competition, inequality, and poverty.

Four

If I Become Rich, Won't Someone Else Become Poor?

"Worrying," says Mark Steyn, "is the way the responsible citizen of an advanced society demonstrates his virtue: he feels good about feeling bad."[1] That's often true. But some things seem genuinely worrisome. Rich Christians (who by global standards probably include almost everyone reading this book) often worry about the gap between the rich and the poor. Our affluence, which is unprecedented by historical standards, makes us see the poverty of others as a moral outrage. When everyone is poor, it's still bad—like disease or earthquakes—but we don't feel guilty about it. It doesn't mean society is unjust. It's when we're prosperous while others are not that we think someone's done something wrong. "The scandal of widespread, persistent poverty in this rich nation" must be called "by its real names: moral failure, unacceptable injustice." That's how a recent gathering of American Christians put it.[2]

In the same vein, after the U.S. Senate voted to raise the minimum wage in February 2007, the Reverend Jim Wallis announced that "God hates inequality."[3] Ron Sider's book *Rich Christians in an Age of Hunger* makes the same point.

Winston Churchill summed up the dilemma with characteristic wit: "The inherent vice of capitalism is the unequal sharing of blessings; the inherent virtue of socialism is the equal sharing of miseries." Most of us know perfectly well that socialist solutions are worse than the disease. But we're still left with capitalism's unequal sharing of blessings. If socialism isn't the answer, what is?

If you look deeper at the problem, you find that for many the real concern is not that some are rich while others are poor, but rather the belief that one causes the other. "The problem with our international global economy," argues Bishop Thomas Gumbleton, "is that the wealth of the world goes from the poor to the rich. The rich get richer and richer. The poor get poorer and poorer."[4] Pope John Paul II, who was no socialist, said similar things in a 1998 sermon in Havana, Cuba:

> Various places are witnessing the resurgence of a certain capitalist neoliberalism which subordinates the human person to blind market forces and conditions the development of peoples on those forces. . . . In the international community, we thus see a small number of countries growing exceedingly rich at the cost of the increasing impoverishment of a great number of other countries; as a result the wealthy grow ever wealthier, while the poor grow ever poorer.[5]

Complaints about the growing gap between rich and poor pop up everywhere.

For decades, Latin American theologian Gustavo Gutierrez has looked around at the poverty in Latin America and concluded that Latin America is poor, in part, because North America is rich: "The poor, dominated nations keep falling behind; the gap continues to grow."[6] It's as if the United States sucks the wealth out of Nicaragua, El Salvador, and Peru and leaves poor Latin Americans without food or houses or land.

So the outrage isn't just that some people have a lot more than others. It's that those who have more seem to have caused others to have less. Even if free trade is a win-win in some abstract sense, "complex developments . . . enable increasingly smaller numbers of people to acquire an ever larger share of the wealth."

Ron Sider continues: "Without corrective action, today's global markets appear to create unjust, dangerous extremes between rich and poor."[7] "Corrective action" usually means forcibly redistributing this unequal sharing of blessings, if not by abolishing private property, as in socialism, then by taxing the rich and giving to the poor—the government as Robin Hood.

Gordon Gekko, the greedy stock speculator from the Oliver Stone movie *Wall Street,* boils the dilemma down to basics:

> The richest one percent of this country owns half our country's wealth, five trillion dollars. One third of that comes from hard work, two thirds comes from inheritance, interest on interest accumulating to widows and idiot sons and what I do, stock and real estate speculation. . . . You got ninety percent of the American public out there with little or no net worth. I create nothing. I own.

> Money itself isn't lost or gained; it's simply transferred from one perception to another. Like magic.

But God is concerned about the poor. We're concerned about the poor. So we find the picture that Gekko paints disturbing.

CHERRY PIE

Imagine the total amount of wealth in the world as a big cherry pie with that flaky crisscross crust on top. You can slice the pie up this way or that. You can cut eight or ten or twelve equal slices, or you can cut some slices thinner than others. If you cut one of the slices really thick, though, you'll have to cut the other slices thinner. It's a trade-off. This, we're told, is the trouble with the world today. The rich get monster slabs of pie while the poorest of the poor get little slivers. It's not fair.

This should sound familiar. It's a restatement of the zero-sum-game problem. We debunked one form of the zero-sum myth by showing that free trade is win-win for everyone who plays. But the zero-sum problem shows up again when it comes to how wealth is distributed. If the world's wealth is like that big cherry pie, there's still a dilemma. If somebody gets a big slice, somebody else is going to get a smaller slice. The details are more complicated than we first thought, what with the win-win of free trade and all, but the trading game started with every kid having something that cost the teacher about a buck. Suddenly things don't seem so win-win if one kid gets a twenty-thousand-dollar dune buggy while the kid next to him gets the broken wheel off of a discarded Hot Wheels toy car.

Substitute fabulous wealth for the dune buggy, and a subsistence living on the brink of starvation for the broken toy car, and you have a picture of the inequality of the modern world. There is something worrisome here. Unfortunately, the heart of the matter often lies hidden under a popular myth about human beings and the nature of wealth.

DON'T MIND THE GAP

The cherry pie is actually a pretty good analogy, but for the wrong thing: it completely misrepresents a market economy. What it represents perfectly is the *myth* about the gap between rich and poor in a market economy. It's one of those background beliefs that are rarely questioned because they're rarely noticed. Think of the pie. It's a physical object; it has a pieish size and shape. It can't shrink much, and it certainly can't grow. We don't know where it came from. It's just there on the kitchen counter getting cold. All we can do is divvy it up and eat it.

But that's not how wealth works in a market economy. Wealth isn't just there: it's not a physical object. No one can simply divide it up at will; and above all, the total amount of wealth can *grow over time*. And as anyone with eyes and even a little historical knowledge knows, it has grown over time. It hasn't stayed the same. If we want to understand wealth and poverty, we must discard the icon of a static and uncreated pie.

POVERTY: RELATIVE AND ABSOLUTE

There are different kinds of poverty, such as absolute poverty and relative poverty. I'm "poor" compared with Bill Gates, for instance, but I still have far more than I need. That's relative poverty. There are many Americans who are "poor" compared with me. But compared with most people prior to the twentieth century, they're rich. For instance, a recent study found some surprising facts about Americans classified as "poor" by the U.S. Census Bureau:

- Forty-six percent of all poor households actually own their own homes. The average home owned by persons classified as poor by the Census Bureau is a three-bedroom house with one and a half baths, a garage, and a porch or patio.

- Seventy-six percent of poor households have air conditioning. By contrast, thirty years ago only 36 percent of the entire U.S. population enjoyed air conditioning.

- Only 6 percent of poor households are overcrowded. More than two-thirds have more than two rooms per person.

- The average poor American has more living space than the average person living in Paris, London, Vienna, Athens, or any of several other cities throughout Europe. (These comparisons are to the average citizens in foreign countries, not to those classified as poor.)

- Nearly three-quarters of poor households own a car; 30 percent own two or more cars.

- Ninety-seven percent of poor households have a color television; over half own two or more color televisions.

- Seventy-eight percent have a VCR or DVD player; 62 percent have cable or satellite TV reception.

- Seventy-three percent own microwave ovens, more than half have a stereo, and a third have an automatic dishwasher.[8]

Compared with the American upper class, these Americans are poor. But that's still relative poverty: it's defined by comparison with others rather than on an absolute scale. On the other hand, if someone's starving to death or freezing to death from exposure because they can't afford shelter, they're suffering *absolute* poverty.

In recent years, ministries like World Vision and multinational groups like the U.N. Millennium Campaign have rightly drawn attention to absolute poverty. The Millennium Campaign lists the following staggering examples:

- One-third of deaths—some 18 million a year or fifty thousand per day—are due to poverty-related causes. That's 270 million people since 1990, the majority women and children, roughly equal to the population of the United States.

- Every year more than 10 million children die of hunger and preventable diseases; that's over thirty thousand per day, or one every three seconds.

- Over 1 billion people live on less than one dollar a day, with nearly half the world's population (2.8 billion) living on less than two dollars a day.

- Six hundred million children live in absolute poverty.

- The three richest people in the world control more wealth than all 600 million people living in the world's poorest countries.

- Income per person in the poorest countries in Africa has fallen by a quarter in the last twenty years.

- Eight hundred million people go to bed hungry every day.

- Every year nearly 11 million children die before their fifth birthday.[9]

For the most part, this depressing list describes absolute poverty. Living versus dying is as absolute as it gets. And living on one to two dollars a day anywhere is barely living.

But one of these items is not like the others: "The three richest people in the world control more wealth than all 600 million people living in the world's poorest countries." We already know that these 600 million suffer absolute poverty. But this statistic isn't about that; it's *comparing* three rich guys who "control" as much wealth as the 600 million poorest people. In 2007, Americans Bill Gates and Warren Buffett and Mexican telecom mogul Carlos Slim Helu had more money than the poorest *10 percent* of the world's population.[10] This is the mother of all gaps between rich and poor. If it upsets you, then it's done its job. It's designed to make you react, to feel morally indignant—but not to think.

Let's do some thinking anyway.

What's the point of the comparison? Is it that some people have way too much money? Are we supposed to think that if Bill Gates weren't so rich, the 600 million wouldn't be so poor? Or worse: are we to suspect that Gates somehow extracted that wealth from the 600 million poorest people on the planet?

We should worry about absolute poverty in the world, especially if we're in a position to do something about it. But worrying about gaps between different groups of people is not the same thing as worrying about absolute poverty. In fact, most such gaps are distractions. Complaints about gaps almost always reflect a basic confusion about the nature of wealth.

When we hear that the gap between rich and poor has grown, for instance, we're supposed to think that means that the rich getting richer makes the poor get poorer. But that follows only if the total amount of wealth is static. Then we've got a zero-sum game, where a gain in one place means a loss someplace else. But in a modern market economy, that's not what happens. The total grows over time. For example, imagine you've got a country, Atlantis, with one thousand citizens. Ten years ago, 999 Atlantans had incomes of one thousand dollars, while one fat cat made $1 million a year—a thousand times more than everyone else. What if I told you that in the last ten years the income gap had grown tenfold? You might think that's bad news, if you *assume* that the gap shows the fat cat getting richer at everyone else's expense.

But without any other information, you wouldn't know the economic status of the other 999 Atlantans. For all you know, they may now make ten thousand dollars a year, and the rich guy, $100 million. Sure enough, the gap has grown tenfold. It's also grown in real dollars, from $999,000 to $99,990,000. But the 999 people at the bottom are making ten times more than they were ten years earlier. They're much better off. They just have a smaller percentage of a much bigger pie. That's not that big a deal. Unfortunately, this point is as easy to understand as it is easy to forget.

Notice that the Millennium Campaign stacks the deck in favor of zero-sum logic. The verb betrays the bias. Why does it say the three rich guys "control" wealth? Why doesn't it say they "own" or "earned" or "have created" wealth? Let's try it: "The three richest people in the world have created more wealth than all 600 million people living in the world's poorest countries." That certainly sounds different, doesn't it? Put this way, the statistic no longer automatically triggers moral indignation. But that's how it should be. We shouldn't be troubled about the wealth Gates, Buffett, and Helu have created.[11] We should be troubled that for some reason, 600 million people have individually owned, earned, and created so little economic wealth. Wealth isn't the problem. Poverty is.

To repeat: Even if the gap between rich and poor grows over time, it *doesn't* mean that the poor are getting poorer, because the total amount of wealth may have gone up. The relevant issue is whether the lot of the poor improves over time, not how close they are to the richest member of their society.[12]

From 1947 to 2005, the average income of the richest 20 percent of the U.S. population went up almost every year, from $8,072 in 1947 to $184,500 in 2005 (adjusted for inflation). But this didn't come at the expense of the poor. On the contrary, the real incomes of the poorest 20 percent also went up almost every year, from $1,584 in 1947 to $25,616 in 2005. And all this happened over a period in which the number of American families *doubled*, from about 37 million in 1947 to over 77 million in 2005.[13] In other words, the total amount of wealth went up. The rich didn't get richer by making the poor poorer. And this is to say nothing of the fact that many families climbed up the income ladder over time. The poorest 20 percent of the population is not always made up of the same people. Upward mobility is common.

The same thing is true internationally. To see this visually, go to the illuminating if badly named Web site Gapminder (www .gapminder.org). Gapminder converts boring, opaque statistics into intuitive animations. It allows you to see trends. One such animation uses an x/y plot to show the trends in life expectancy and per capita income from about 1974 to 2005. If you can get on the Internet, before you read further go to www.gapminder .org and click on "Gap Minder World, 2006." Then you can follow along. Words alone don't do justice to the reality.

Every country is represented here on an x/y plot with a color-coded circle. The country's population determines the size of the circle. (That's why India and China look like Jupiter and Saturn while most of the other countries look like little moons.) The y-axis (up and down) shows life expectancy. So the higher up a country is on the plot, the higher its average per capita life expectancy. The x-axis (left to right) shows per capita income. The farther a country is to the right on the plot, the higher its per capita income. Now hit "Play," and watch the circles move through time. Notice the general trend: up and to the right. In other words, per capita income and life expectancy have gone up in many countries in the last thirty years, especially in Europe, Asia, and North America. Total income has increased worldwide.[14]

Unfortunately, some countries buck the trend, especially in sub-Saharan Africa. Rwanda, for instance, wanders all over the place. Life expectancy drops below twenty-five in the early 1990s(!), followed by a plummeting per capita income after that.

This corresponds to the bloodbath between the Hutu and Tutsi tribes during that time. Mass murder has economic and public health consequences that show up on graphs.

Angola, under the influence of communism, wanders around the same spot on the plot for twenty years. And several other African countries, like Nigeria and Ethiopia, move leftward for years in a row. Socialist dictatorships like the Democratic Republic of Congo consistently move left. They restrict economic freedom, and get poorer and poorer. China and India, in contrast, continue a steady glide up and to the right as their economies grow freer. In short, there's no international trend of the rich getting richer by making the poor poorer.

In fact, the percentage of people living in absolute poverty has dropped since 1970. In 1970, the world population was 3.7 billion, and 38 percent (1.4 billion) lived below the absolute poverty line (less than one dollar a day). By 1990, with a world population of 5.3 billion, those languishing in absolute poverty dropped to 26 percent (still about 1.4 billion).[15] In fact, despite puddleglummish reports to the contrary, worldwide, statistics on infant mortality, life expectancy, and poverty have all improved dramatically in the last few decades.[16]

Comparing countries, there is one unmistakable trend: countries with the rule of law and economic freedom prosper over time. Countries without these virtues do not. The annual "Index of Economic Freedom" drives this home. In 2007, booming Hong Kong topped the list, while starving, Stalinist North Korea came in dead last.[17] Those two facts tell you what you need to know. If every country had free markets and the rule of law, every circle on the Gapminder plot would probably be moving up and to the right.

MATTER ISN'T ALL THAT MATTERS

So we know that the total amount of wealth, of capital, increases over time in a market economy. But how is that possible? Isn't wealth made up of stuff like gold, diamonds, oil, and land?

Surely there's only so much of that to go around. If Paul is getting a big fat share of these goodies, somehow, somewhere, somebody's got to be robbing Peter, right? Wrong.

Although we link wealth with material possessions—with stuff—the essence of wealth, even though it involves matter, is immaterial. It's only a slight exaggeration to say that wealth isn't about stuff; it's about us. To believe otherwise is to fall for the most pernicious economic myth of all—the materialist myth. Materialism as a broad idea isn't the belief that matter is good, or useful, or real. Every Christian should believe that. You find it in the creation story where God says everything he made was good. And even after the fall of Adam and Eve into sin, you have God becoming flesh and dwelling among us. Materialism doesn't say that matter is good. Materialism doesn't even have room for the categories of good and evil. Materialism is the belief that matter is the fundamental reality, that everything derives from matter, that matter is all that matters.

As a worldview, it was summed up in the old PBS series *Cosmos* by the show's host, astronomer Carl Sagan, who informed viewers that "the cosmos is all that is, or every was, or ever will be." Sagan's materialist credo contradicts the Christian view that in the beginning, God created the heavens and the Earth—that is, the entire material universe. Just from that you know that at least one thing exists beyond the cosmos: the Creator. No half-thinking Christian will be a strict Sagan-variety materialist. Nevertheless, materialism is in the air we breathe, so it shows up in our thinking without our even realizing it. It especially contaminates our understanding of wealth.

Myth no. 4: The Materialist Myth (believing that wealth isn't created, it's simply transferred)

Recall something discussed earlier. For centuries, economists thought the economic value of a good or service was something we can touch or see—either something in the matter itself or in the amount of labor it took to produce it. Instead, economic value is about how we value goods and services. In a competitive market, prices are packets that contain information about how much a good or service is valued, and how scarce it is. At the very heart of economics, then, is a reality that exists not in material objects but in our individual and collective minds.

Just as this applies to goods and services, it also applies to money. Money is a socially constructed reality that represents wealth or capital. We can use units of currency, like dollars, to measure how much capital someone has accumulated. But money isn't capital itself. Peruvian economist Hernando de Soto puts it this way:

> Capital is now confused with money, which is only one of the many forms in which it travels. It is always easier to remember a difficult concept in one of its tangible manifestations than in its essence. The mind wraps itself around "money" more easily than "capital." But it is a mistake to assume that money is what finally fixes capital. Money facilitates transactions, allowing us to buy and sell things, but it is not itself the progenitor of additional production.[18]

Money has value only if trading partners believe it has value. This is why currency quickly becomes stove fuel when people stop trusting it.

OK, but what about property? Maybe my intellectual property and diligence and ideas are immaterial, but what about land? Land has a location, boundaries, dirt. That's matter if anything is. Yes, of course plots of land on the earth's surface are made of matter. But that's not what makes them *property*. "Property is not really part of the physical world," argues de Soto, "its natural habitat is legal and economic. Property is about invisible things."[19] Although squatters may be able to extract some value out of land they occupy by sleeping there or by planting cash crops that they can quickly harvest to eat or to sell, that land is not their property unless it is represented as part of a formal system that is widely seen as legitimate. "Property is not the assets themselves, but a consensus between people as to how those assets are held, used, and exchanged," de Soto argues.[20]

If the same squatters acquire a binding legal title representing the same land, what de Soto calls "pure concept," then they own property. It's property not because it's made of dirt, but because it's represented by a title that reflects an underlying social reality. "Remember, it is not your own mind that gives you exclusive rights over a specific asset, but other minds thinking about your rights in the same way you do. These minds vitally need each other to protect and control their assets," he explains.[21] These aren't social fictions. They're *realities*. In economics, de Soto points out, despite vulgar appearances, "not everything that is real and useful is tangible and visible."[22]

This system of representation allows land to *become* property, to *become* capital, to be used as collateral for loans, to be compared with and combined with other assets. Owners can then tap their property's potential. Recently a group of economists headed by Rafael Di Tella of the Harvard Business School studied a squatter settlement in Argentina in which some squatters acquired legal right to their land while their neighbors did not. They discovered that owning property changes the way people look at the world. The property owners were more likely to trust others, and to believe that individuals could achieve something good if they worked hard.[23] This makes sense. Property owners have a vested interest in the health and stability of the social order that makes their property possible. They have something to gain and something to lose. They treat the same plot of land differently.

For decades, there has been a close correlation between the wealth of a country's citizens and the strength of that country's property laws. In general, the more a country (which includes the government and its citizens) protects private property, the more prosperous the citizens of that country will be.

**Correlation of Legal Property
Protection and GDP per Capita**

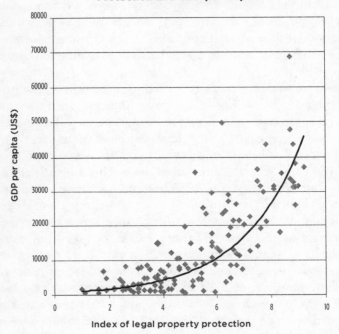

Figure 2. James D. Gwartney and Robert A. Lawson, *Economic Freedom of the World: 2006 Annual Report; CIA World Fact Book,* https://www.cia.gov/cia/publications/factbook/rankorder/2004rank.html.

But this is only the beginning. All of these immaterial realities allow us to *create* wealth that didn't previously exist. When you first hear this, it sounds strange, like a pyramid scheme or some New Age quackery. But that's only because we're used to thinking like materialists when it comes to wealth. Ironically, even many Christians think this way. That's a mistake, since the Bible describes us as creators from the very beginning.

CO-CREATORS

The epic of creation in Genesis 1 describes God creating everything in six days: "In the beginning, God created the heavens and the earth." God, like human beings, has a workweek. God, like us, works during the day and rests at night. The six days of his creation are separated by "an evening and a morning." Unlike earthly days, however, which are marked by the sun's apparent movement across the sky as the earth rotates on its axis, God creates and then separates the light and darkness that mark his days. He fixes his own workdays. God calls forth everything freely and effortlessly, simply by saying, "Let there be . . . light . . . waters . . . living creatures." And at the end of each day, Genesis says, "God saw that it was good."

The sixth day starts like every other, with God saying, "Let the earth bring forth . . ."—in this case, ". . . living creatures of every kind." But on this day, there's an encore. Rather than simply saying, "Let there be," God now speaks to himself in the plural: "Let us make humankind in our image, according to our likeness; and let them have dominion over the fish of the sea, and over the birds of the air, and over the cattle, and over all the wild animals of the earth, and over every creeping thing that creeps upon the earth." Then God blesses the man and woman, saying: "Be fruitful and multiply, and fill the earth and subdue it; have dominion over the fish of the sea and over the birds of the air and over every living thing that moves upon the earth." While God saw on previous days that everything he had made was good, he now sees that everything he has made is "very good." Human beings aren't God's only purpose in creation; but we're his crowning achievement.

Theologians have often equated the image of God—the *imago dei*—with one part of our nature, such as our reason or our soul; but the biblical text simply says that human beings—all of us, male and female—are true icons of God. The text is sparse on details, but it flatly contradicts what Israel's neighbors believed. They saw their kings or their idolatrous images as representing the divine. The Egyptian pharaohs represented the gods, for instance, but ordinary Egyptians did not.[24] Genesis, in contrast, says that each of us is made in God's image.

God is king over the heavens and the earth, and appoints us to have dominion as kings and queens over his creation. He creates simply by calling things into existence. He then commands us to create according to our power, to "be fruitful and multiply, and fill the earth and subdue it." God is the Creator, and human beings, his true representatives, are co-creators. That doesn't mean we're creators as God is. To avoid that suggestion, J. R. R. Tolkien preferred the term "sub-creators." But whatever we call it, the key point is that our creativity comes from God. It doesn't compete with him. As Creator, God has made us with the awesome power and responsibility to *create*.

Where Genesis 1 gives the cosmic overview, Genesis 2 narrows the focus. God fashions humankind—man—from the ground, breathes into him the breath of life, and puts him in a garden "to till it and keep it."[25] We aren't ethereal angels or wispy spirits. We're made of dirt, and we're made to work with dirt—and yet we have in us the very breath of God. We're drawn from the ground but still transcend it. Dirt, as God created it, isn't dirty. And working with dirt, as God created it, doesn't make us dirty. Work itself is part of God's original blessing, not his curse after the fall. The way in which we work, then, should reflect the fact that we are a unity of matter and spirit, of heaven and earth, neither pack animals nor angels.

These chapters of Genesis aren't economic texts, but they still cast light on the most important truth of economics: not only do we create wealth, but, in the right circumstances, we create more and more wealth. Wealth, rightly applied, begets more wealth.

We see this clearly when we survey human history. When we first arrived on the scene, most humans were mostly hunters and gatherers. A few thousand years ago in the Fertile Crescent (modern Iraq), some humans began to domesticate sheep and cattle and to cultivate plants like wheat and barley.[26] Farmers start with what God has provided—seeds, land, rain, animals— and enhance it.

The word "capital" derives from the Latin word *caput,* meaning "head." In Medieval Latin, *caput* referred to head of cattle or livestock.[27] And for most of human history, that made sense. For thousands of years, almost everyone worked the land as farmers or shepherds. Even today, economists list the three factors of production as land, labor, and capital. After people began developing ways to store food and irrigate land, cities started to spring up, with growing populations not tied directly to agriculture. From cities grew much larger civilizations, like the Egyptians, whom Joseph advised to store grain for a seven-year famine. As in ancient Egypt, slave labor was common everywhere for centuries. It was eventually replaced by feudalism, especially in Europe, which then gave way to widespread private-property rights in the West.

From the Middle Ages until the present, stone and wood implements were gradually replaced by metal plows and wheeled carts pulled by oxen and then horses. Technology such as seed drills, reapers, steam power, tractors, and combines transformed farming, so that fewer people could produce more with less land. Everything from pest control to better cultivation made farms even more fecund. Farms sprang up in areas that were previously arid. And yet, as recently as 1900, four-fifths of the world's population still lived on farms.

By the 1970s, though, only half the world's population was farming. That's the worldwide *average*, however. In much of the world, people still engage in primitive subsistence farming, as they have for centuries. That means they grow just enough to feed themselves, if they're lucky. In the United States and other societies with solid private-property laws and high technology, however, a much smaller percentage of people make their livelihood by farming. Only 1.9 percent of the American population now lives on farms, and yet they produce enough not only to feed the American population but to export abroad. In principle, the state of California could now grow enough to feed the planet. We have reached a unique moment in history.

A few years ago, my uncle had to fly to Seattle, where I lived, for a medical procedure. He had spent almost his entire life in the sparsely populated Texas Panhandle, which is covered mostly with farms and ranches. Near the end of his flight, he saw the miles and miles of mountains, forests, and water that surround

Seattle, and was startled to see the thriving Emerald City spring up at what seemed to be the edge of the earth, with nary a farm in sight. After he landed, he asked me: "What do all these people *do*?" Good question. Most of us don't ask it only because we take our modern economy for granted.

When farm output increased, many people were freed to spend their time in other pursuits. Better farming opened the way to developments in art, music, philosophy, literature, and all sorts of new technologies. Leisure, it's often said, is the basis of culture. Without plows and cultivated grain, we might never have had Moses, Jerusalem, Aristotle, and Rome. From mastering fire and inventing the wheel, to learning to draw metal ores from the earth with mining and smelting, to harnessing the energy of rivers with water mills, our ancestors slowly transformed the material world around them to create new wealth. From the pencil to the iPod, when we look at technology, we *see* created wealth. But technology is only the tip of the iceberg, the tangible tip of an immaterial reality.

In the second chapter of Genesis, right after God puts the man in the Garden, even before he creates the man's helper, God brings the animals and birds "to the man to see what he would call them; and whatever the man called every living creature, that was its name." God created the animals, but he didn't give them names. He gave that honor to the man. From the beginning, we've been signifying beings. Unlike apes in the jungle, we devise abstract ways of representing the world. We speak, we name, we write, we draw, we tell bad jokes, we invent clocks—"time-keeping" devices—to represent the passage of time, we print money to represent assets, we develop simple and complex mathematics to count our cows and predict the orbits of the planets, we develop laws and procedures to protect ourselves and our property. When we do this, we magnify our capacity to communicate, to cooperate, and to create new wealth.

Though the word *capital* originally referred to cattle, it's also reminiscent of the Latin word for head, *caput*.[28] That's much more fitting for modern economies, in which the mind predominates.

Hernando de Soto points out that property laws make capital "mind friendly," since they allow physical assets like land to take on a parallel existence in the realm of representation. But this is only the beginning of an information economy. Most of the vast new wealth created in our economy relies on symbols, information: calculations of interest in a bank, modulations of radio waves and microwaves in the air, and laser beams riding tiny tubes of glass—fiber-optic cables—across the bottom of the ocean.

The modern computer revolution was made possible by an ingenious way of encoding information, using only two characters—the humble 0 and 1. We have woven this symbolic system into matter in such complex ways that some conspiracy theorists think we learned it from advanced extraterrestrials. We now use virtually free material, silica, like the sand on the beach, to make computer chips and fiber-optic cables—the brain and nervous system of a global information revolution. A fiber-optic cable made from sixty pounds of cheap sand can carry one thousand times more information than a cable made from two thousand pounds of expensive copper.[29] God made sand; he let us transform it into chips and cables. We encode information in matter—we *inform* matter—by writing with a stick in sand, and by impressing 0s and 1s magnetically onto a hard drive. The cost of storing and transferring information goes down over time. Information, remember, isn't matter. Information comes from minds.

Even more startling is that over time, we can create more and more wealth with less and less matter. It took perhaps six thousand years to go from the first farms to the invention of the wheel. In contrast, everything from indoor plumbing to telephones, lightbulbs, cars, planes, rockets, computers, MRIs, and antibiotics was invented in the last four generations. In advanced societies, the creation of wealth is now following an exponential trajectory in which creativity dominates mere matter. Futurist Ray Kurzweil has popularized this idea with his "law of accelerating returns."[30] Although Kurzweil over-extends the argument, he's still onto something profound. We now measure everything from increases in wealth to changes in technology in months rather than centuries. One way to see this is to consider the number of patents granted in the United States since 1870.

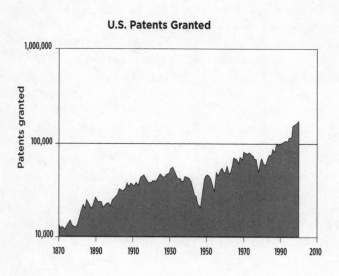

U.S. Patents Granted

Figure 3. Courtesy of Ray Kurzweil and Kurzweil Technologies, Inc., http://www.kurzweilai.net/index.html?flash=1.

At the very end of the nineteenth century Charles H. Duell, commissioner of the U.S. Office of Patents, announced, "Everything that can be invented has been invented." It would have been hard for him to be more wrong. Not only did inventions keep coming; they came at a faster rate. Around 1870, some 15,000 patents were granted each year in the United States. The annual total now exceeds 150,000. That's a tenfold increase!

Another popular example of accelerating returns is Moore's Law, named for Intel cofounder Gordon Moore. In 1965, Moore observed that the number of transistors in an integrated circuit *doubles* every twenty-four months while the cost for the new circuit stays the same. This can be extended across the entire twentieth century by describing it as one stage of a trend of increasing computer calculations per second per one thousand dollars. This takes account of the fact that computer technology has changed several times in the last century and can be expected to change again in the century ahead.

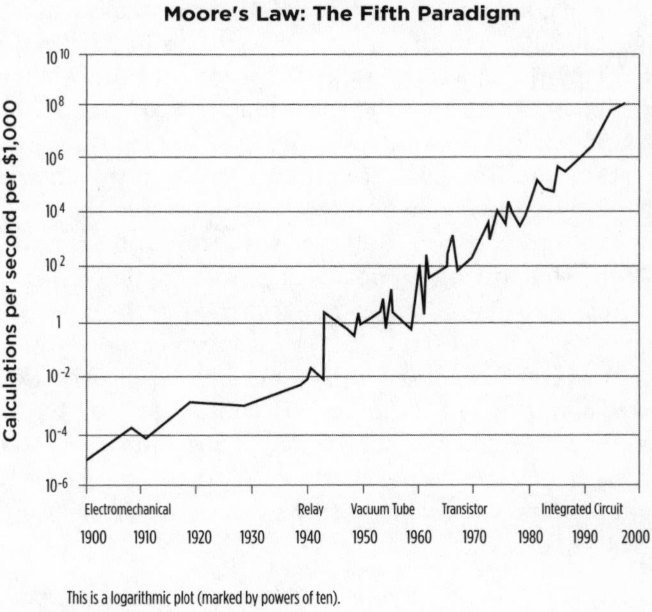

Moore's Law: The Fifth Paradigm

Calculations per second per $1,000

10^{10}, 10^8, 10^6, 10^4, 10^2, 1, 10^{-2}, 10^{-4}, 10^{-6}

Electromechanical   Relay   Vacuum Tube   Transistor   Integrated Circuit

1900 1910 1920 1930 1940 1950 1960 1970 1980 1990 2000

This is a logarithmic plot (marked by powers of ten).

Figure 4. Courtesy of Ray Kurzweil and Kurzweil Technologies, Inc., http://www.kurzweilai.net/index.html?flash=1.

Notice that these are real trends, not speculations. Notice also that they're happening in the high-tech, information-dense parts of the economy—the places where mind predominates.

It's a mistake to refer to such trends as "laws," since they're not inevitable. A plague, nuclear war, or stupid economic policy could stop them in their tracks. Besides, there's no way of knowing whether these growth *rates* will continue indefinitely. But since we are creators, we have every reason to expect that we can continue to create new wealth. That's because, contrary to untutored common sense, new wealth comes not from matter alone, but from how we represent, inform, and transform matter. If we miss that fact, we miss the most profound truth of economics—a truth Christians should have expected all along.

That's the good news. The bad news is that wealth isn't created automatically, like a metal ball falling to the ground when dropped from the Leaning Tower of Pisa. Wealth is created only in the right circumstances. There are still far too many places in the world that are economically stagnant, where life goes on as it did hundreds or even thousands of years ago. In much of Central America and Africa there are tiny pockets of wealth, but most of the population still lives in abject poverty. And even in the most prosperous societies, like the United States, pockets of poverty remain.

If trends continue, the gap between rich and poor will grow *exponentially* wider, even as the lot of the poor slowly improves. Except in the case of theft, this won't be because the rich have extracted wealth from the poor.[31] It will be because wealth creation is on a trajectory of accelerated returns in some places and is scarcely being created in other places. Again, the gap is not the problem. The problem is that some places aren't creating much wealth.

WITH *EQUALITY* AND JUSTICE FOR ALL?

OK. Since wealth is *created,* the total amount of wealth isn't fixed. My wealth needn't cause your poverty. But there's still the problem of inequality. Remember Churchill's words: "The inherent vice of capitalism is the unequal sharing of blessings." That still bugs us. We Americans believe in equality. It took a couple of centuries to work this out consistently, what with slavery and arbitrary restrictions on voting; but America and equality have gone together from the beginning. Thomas Jefferson minced no words when he wrote the lines that declared the independence of those first thirteen colonies from the British Empire: "We hold these truths to be self-evident, that all men are created equal." These are strong words, since self-evidence is an honor normally reserved for the truths of math and logic, which we recognize the second we understand them.

Most cultures agree that one plus one equals two; but historically, few would have agreed that human beings are equal. In India, people were assigned to different castes, with the Brahmans at the top and untouchables at the bottom. The ancient Greeks thought citizens of city-states like Athens were worthy of certain rights and privileges. But they saw no problem with slavery, as long as the slaves were "barbarians."

In fact, slavery was as common as the cold until the British Empire and then other Western nations abolished it in the nineteenth century.[32] Slavery didn't become illegal worldwide until *1970*, however, when it was finally outlawed on the Arabian Peninsula. Even so, it's still rampant in Africa and Asia. And the coercive labor in communist regimes like North Korea is really slavery with the state holding the monopoly.

We know slavery is wrong; but can you blame people who doubt human equality? After all, look around. To judge from appearances, inequality is the plain truth. Some people are tall. Some are short. Some are fat. Some thin. Some smart. Some not so smart. Some hard working. Some lazy or at least easily distracted. Some kind. Some curmudgeonly or just mean. Some natural athletes. Some happy just to be able to walk and chew gum at the same time. Is there any belief more in conflict with the evidence before our eyes than this belief that we're all equal? And we've seen what happens when the state tries to *make* everyone equal: misery, starvation, death. So maybe we should just jettison this naive belief and face reality squarely.

But not so fast. Recall Jefferson's words. He knew the myriad differences between us as well as anyone. He based our equality, however, not on a comparison, but on our common origin: "We hold these truths to be self-evident, that all men are created equal, that they are endowed by the Creator with certain unalienable rights." We are all equally entitled to certain rights because we're all humans made by God and standing in a special relationship to God, over and above that enjoyed by other creatures. Of course, this was a biblical idea before it was an American idea. According to the Bible, we're equal because we are, each of us, created in God's image. As the prophet Malachi declared: "Have we not all one father? Has not one God created us?" (Mal. 2:10).[33] Jesus modeled this by treating religious teachers, tax collectors, prostitutes, and Samaritan women with dignity. Because one God has created us all in his image, we all deserve to be treated as equals before the law. The same rules should apply to the rich and the poor, the weak and the strong, the winners and the losers.

So what do we do with run-of-the-mill differences? Inequality in most areas doesn't cause much consternation. Van Cliburn can play the piano much better than I can. Michael Jordan can play basketball and baseball far better than I can. Pavarotti could sing better than practically everyone. Does that prove God is unjust? Is God obliged to distribute his blessings equally? That's not the impression you get from the Bible. Think of Jesus's parable of the talents (Matt. 25:14–30), in which a man gave his slaves different amounts. One gets five talents, another two, and another one. Later he judges them by what they've done.

Or think of another parable, of a landowner who goes out early in the morning to hire workers for his vineyard. The landowner agrees to pay the going rate—a denarius—to the first workers he hires in the morning. But he ends up hiring more workers at nine o'clock, then again at noon and three o'clock. He even hires a few stragglers at five o'clock. At the end of the day, the landowner pays all the workers the same amount. The ones who work only one hour get the same as the workers who labor all day. Ah-hah! Everyone is treated equally, right? But the early birds in the parable don't see it that way. When they get wind of this, they grumble that they didn't get more. But the landowner rebuffs one of them: "Friend, I am doing you no wrong; did you not agree with me for the usual daily wage? Take what belongs to you and go; I choose to give to this last the same as I give to you. Am I not allowed to do what I choose with what belongs to me? Or are you envious because I am generous?" (Matt. 20:13–15).

Of course, this parable isn't primarily an economic lesson. Jesus begins the parable by saying: "The kingdom of heaven is like . . ." He's illustrating a mystery about the kingdom of God: when it comes in its fullness, "the last will be first, and the first will be last." But that doesn't make the parable any less relevant to our question. The landowner represents God, so clearly Jesus sees his actions as just. If the landowner had not paid the first workers what he promised, he would have been guilty of theft. *That's* injustice. As it is, the landowner keeps his promises. He just pays each a different rate per hour. But so what? That's his choice to make. Instead of being pleased for receiving what they were promised, the early risers envy the others for what they have received. We all tend to do that—to link inequality of outcome or opportunity with injustice. But they're not the same thing.

When Jim Wallis announces that God hates inequality, he's not quoting the Bible. The Bible doesn't say that anywhere. Jesus tells his inner circle of followers that they will sit on twelve thrones (Matt. 19:28). He speaks of the least now being the greatest in the kingdom of God. He tells his disciples that those who are humble like children will be greatest in the kingdom of God (Matt. 18:4). If absolute equality doesn't apply to God's kingdom—the very standard of justice—why do we think it should apply to human society? If God can be just without sharing his blessings equally, might the same be true of a society?

Ron Sider agrees that we shouldn't try to make everyone's income equal. Still, he argues that extreme disparities in wealth are dangerous . . . to democracy. Wealth is power, after all, and as Lord Acton famously observed, power tends to corrupt, because all human beings are fallen. "Concentrated wealth," Sider argues, "equals concentrated power."[34] To avoid concentrated power, the American founders distributed the powers in the federal government among the congress, the president, and the courts. Sider argues that the same insight that led the founders to separate political power should lead us to separate economic power by preventing wealth from getting concentrated in the hands of the few. Otherwise, the wealthy will wield far too much power over the political process.

This sounds plausible, but think about it. How do you prevent people from acquiring vast wealth? Government could work to keep everyone poor. Governments can't do everything, but history clearly shows that governments are more than capable of creating widespread poverty. But no sane person would recommend that policy. What Sider and others advise instead is that the government tax the wealthy disproportionately to prevent them from getting too rich, and then redistribute that wealth to poorer members of society.

Normally, it's unjust to treat citizens differently based on their income, let alone to take one person's property and give it to others. Moreover, we already have a progressive income tax. In 2004, the top 1 percent of income earners had a 19 percent share of the total income but paid 36.89 percent of the income taxes, and the top 50 percent of income earners paid 96.7 percent of the income taxes.[35] But let's set these points aside. Remember, Sider argues that we should "redistribute" to *prevent power from getting too concentrated* in the hands of the few. To do that, he recommends that the government, which has the unique coercive power to make and enforce the laws, should confiscate the legal wealth of some citizens and give it to others. Sider's remedy for preventing a concentration of power among citizens is to pit the interests of some citizens against others, and to concentrate more power in the state. But the federal government's budget is much larger than the net worth of even the wealthiest Americans, and its power far more vast. So if we're concerned about concentrated power, why on earth would we want to hand more power over to the most powerful entity in human history, the U.S. government? That doesn't make any sense. The cure is worse than the disease.

Policies like those Sider recommends led Scottish statesman Alexander Fraser Tyler (1742–1813) to conclude that "a democracy cannot exist as a permanent form of government." Instead, he argued:

It can only exist until the voters discover that they can vote themselves largess from the public treasury. From that time on the majority always votes for the candidates promising the most benefits from the public treasury, with the results that a democracy always collapses over loose fiscal policy, always followed by a dictatorship.

How long can a society exist once a majority discovers it can vote to have the state confiscate the wealth of a minority and give it to them? I don't know. Perhaps over time this is unavoidable; but it's beyond me how a Christian ethicist can advocate the policy outright.

Of course, Sider is right that the wealthy tend to wield greater influence in politics than those of more modest means; but in open societies like the United States, the "wealthy" aren't a uniform voting block. They're individuals who disagree on pretty much everything, especially politics. George Soros and Teresa Heinz Kerry pour millions into left-wing causes. T. Boone Pickens and Rich DeVos support conservative causes. Some corporate CEOs are Democrats. Some are Republicans. Others are independents all across the political spectrum. Their power is checked by the simple fact that they don't all think alike.

Bill Gates told the Harvard graduating class of 2007 that economic inequality was a moral outrage and that "reducing inequity is the highest human achievement."[36] Jim Wallis has complained that the "great crisis of American democracy today is the division of wealth." Nonsense. If that were true, I would be right to seethe in anger because I have less than Bill Gates. That's envy, not justice. Hand wringing about gaps and inequality does little more than encourage perverse schemes like those favored by Wallis and Sider, where the state attempts to play the role of unerring distributor of wealth. No such government institution exists, and wherever it has taken up the task with great zeal, it has not eliminated poverty but spread it like a virus.

In the kingdom of God, there will still be inequality, but there won't be poverty. Between the Garden and the present day, however, poverty has been the norm. And Jesus tells us that we will always have some poor among us. Until a few centuries ago, however, most Europeans were as dirt-poor as many in the third world are today. Historically, our current pockets of prosperity are highly unusual. We rightly see poverty as a problem, just as disease is a problem. But the problem isn't that some people are rich and some are poor, any more than the problem of disease is that some people are healthy. The problem is quite simply that some are poor. If we want to bask in the wasteful heat of self-righteous moral indignation, then by all means, let's keep blathering on about income gaps. But if we really want to help the poor, we need to get our eyes off decoys and focus on the real problem—*poverty*—and its only known solution: creating wealth.

Five

Isn't Capitalism Based on Greed?

You may recall the 1987 movie *Wall Street* and the ruthless corporate raider named Gordon Gekko. Gekko is famous for his defense of selfishness: "Greed . . . is good," he tells a young broker. "Greed is right. Greed works. Greed clarifies, cuts through and captures the essence of the evolutionary spirit. Greed, in all its forms . . . has marked the upward surge of mankind." Gekko embodies the enduring stereotype of the greedy businessman, so enduring that 20th Century-Fox decided to produce a sequel some twenty years after Gekko first appeared on screen.

But truth is sometimes stranger than fiction. Gekko was modeled on the real-life investor Ivan Boesky, who immortalized himself in 1986 when he addressed graduates at the University of California at Berkeley in a commencement speech: "Greed is all right, by the way," he asserted. "I think greed is healthy. You can be greedy and still feel good about yourself." (Later Boesky, like Gekko in *Wall Street,* went to prison for insider trading.)

Gekko is a caricature; and yet many of capitalism's champions admire him. Michael Douglas, who played Gekko, told the *New York Times* he would rather not have "one more drunken Wall Street broker come up to me and say, 'You're the man!'"[1] ABC's John Stossel once built a TV special around the question, Is greed good? These brokers would probably answer yes without so much as a wink or a nuance.

Taken at face value, all this greed-is-good rhetoric is a big fat nonstarter for Christians. It's not like the Bible and the Christian tradition have been vague on the matter. Greed, or "avarice," is one of the seven deadly sins, and the Bible has nothing good to say about it. In the Gospels, when Jesus was asked to settle an inheritance dispute, he responded: "Take care! Be on your guard against all kinds of greed; for one's life does not consist in the abundance of possessions" (Luke 12:15). The Tenth Commandment says, "Do not covet," which no doubt applies to greed as well. Jesus includes greed with murder and adultery in a long list of sins (Mark 7:21–22). Paul tells the Ephesians that no greedy person—"that is, an idolater," he explains—will inherit the kingdom of God (Eph. 5:5). This is a representative sample of the dozens of biblical passages condemning greed. No matter how many seminary degrees you have, there's just no explaining these away.

A friend recently heard a sermon in a conservative evangelical megachurch in Grand Rapids, Michigan, about the most prevalent anti-Christian philosophies. The preacher condemned humanism, hedonism, postmodernism, and . . . capitalism. It quickly became clear that by "capitalism" the preacher meant greed. It's hard to blame this pastor, since prominent evangelicals like Tony Campolo complain that capitalism is based on the "greed principle."[2] And it's hard to blame Campolo, since even many *fans* of capitalism agree.

Despite this agreement from friends and foes alike, however, capitalism is not based on greed. The critics attack a false stereotype; the champions defend it. The truth is much more interesting, and much more inspiring. As always, to get this, you just have to think.

BAD FOR THE SOUL, BUT GOOD FOR SOCIETY

It all started with bees. Well, with a story about bees, anyway. In 1705, a Dutchman named Bernard de Mandeville wrote a long poem called *The Grumbling Hive; or, Knaves Turn'd Honest.* Nobody noticed it. So in 1714, he republished it in *The Fable of the Bees,* a book that included a lengthy commentary explaining that the poem was a metaphor for English society. Following convention, the book had a title you could remember, followed by a boring subtitle that explained what the book was really about: *Private Vices, Publick Benefits*. Mandeville saw humans and bees as little more than bundles of vicious passions. *The Fable of the Bees* reflected that belief.

The poem describes a "Spacious Hive well stock'd with Bees." Like modern society, the hive has a division of labor. Different bees do different tasks, but they all have the same motivation—vice:

> Thus every Part was full of Vice,
> Yet the whole Mass a Paradise.

The poem describes avarice, pride, and vanity as producing great wealth for the hive:

> Their Crimes conspired to make 'em Great;
> And Vertue, who from Politicks
> Had learn'd a Thousand cunning Tricks,
> Was, by their happy Influence,
> Made Friends with Vice: And ever since
> The Worst of all the Multitude
> Did something for the common Good.

Vice "nurses ingenuity," making possible a vast and prosperous population of bees. So observed from the outside, everything is splendid; but from the inside, it looks much worse: the bees in the hive grumble at the lack of virtue around them. They gripe so incessantly that Jove eventually gives them what they ask for. Honesty and virtue now fill the hive. And everything falls apart:

> For many Thousand Bees were lost.
> Hard'ned with Toils, and Exercise They counted Ease it self a
>   Vice;
> Which so improved their Temperance;
> That, to avoid Extravagance,

They flew into a hollow Tree,
Blest with Content and Honesty.[3]

So the bees' virtuous actions led to disaster, whereas the individual acts of evil had led to social good. As Mandeville explains:

> I flatter my self to have demonstrated that, neither the Friendly Qualities and kind Affections that are natural to Man, nor the real Virtues he is capable of acquiring by Reason and Self-Denial, are the Foundation of Society; but that what we call Evil in the World . . . is the grand Principle that makes us sociable Creatures . . . the Life and Support of all Trades and Employments without Exception: That . . . the Moment Evil ceases, the Society must be spoiled, if not totally dissolved.[4]

Thus interpreted, the book caused a major scandal in polite society: picture powdered European women in impractical hoop skirts fainting upon hearing the name Mandeville on the street.

Taken literally, Mandeville's claim is ridiculous. Good doesn't come from evil. Virtue isn't born from vice. Virtue doesn't destroy society. A place where everyone pursued their most vicious instincts would resemble *The Lord of the Flies*, not *Bees in Paradise*. At best, Mandeville was guilty of hyperbole. At worst, he got something horribly wrong. But he did get one thing right: individual acts of evil can lead to good social outcomes.

We're familiar with the Law of Unintended Consequences. It usually involves good intentions gone awry. For instance, let's say Senator Ted Kennedy reads an article in *Sojourners* claiming that social justice requires a higher minimum wage.[5] So he sponsors legislation to raise the federal minimum wage to eighty dollars an hour. Improbably, his bill makes its way through both houses of Congress and gets signed by the president. Once it's the law of the land, that policy will have an effect: before long, no one whose labor is worth less than eighty dollars an hour will have a legal job. It doesn't make a dime's worth of difference whether Ted Kennedy sponsored the bill because he wanted to help poor people, union workers in Boston, or his chances of winning the Nobel Peace Prize. Good intentions don't always yield good results.

The dark side of that truth is that bad intentions don't always yield bad results. The apostle Paul once delighted that some were preaching the gospel out of envy of him. He delighted not in the envy but in the preaching. Even private sinful acts may lead to a social good. In 1978, Bernie Marcus and Arthur Blank were fired as executives of Handy Dan Home Improvement Centers. The next day, they began conspiring to exact revenge on their former employer—not by theft or terrorism, but by starting a competitor.[6] Today, the company they started, Home Depot, employs more than 355,000 people in some twenty-two hundred stores around the world. Marcus's and Blank's initial motives were morally bad for them and were financially bad for Handy Dan: it's gone the way of dinosaurs and dodo birds. But Marcus and Blank created hundreds of thousands of jobs while helping dishwashers, janitors, middle managers, and everyone in between to improve their homes. It doesn't follow that you should sin so that good may result, or that good outcomes justify immoral motives. The point is that private motives don't determine outcomes. If you get this, you'll soon see why capitalism doesn't equal greed.

THE VIRTUE OF SELFISHNESS?

After Mandeville came the Scottish philosopher Adam Smith, who in 1776 wrote the most famous book in the history of economics, *An Inquiry into the Nature and Causes of the Wealth of Nations*. Though the book is long on pages and detail, Smith's basic purpose was simple. He wanted to defend what he called the natural system of liberty: rule of law, unobtrusive government, private property, specialization of labor, and free trade. To prosper, a society needed "little else," he said, "but peace, easy taxes, and a tolerable administration of justice."[7] Smith is a patron saint of capitalism. But far from sucking up to the business class, he famously said that "people of the same trade seldom meet together, even for merriment and diversion, but the conversation ends in a conspiracy against the public, or in some contrivance to raise prices."[8] Not exactly a glowing endorsement.

Smith never attributed the happy outcomes of trade and business to the virtues of businesspeople. "It is not from the benevolence of the butcher, the brewer, or the baker," he wrote, only to be quoted by every economics textbook ever written, "that we expect our dinner, but from their regard to their own interest."[9] Nevertheless, through the invisible hand of the market, individuals will "promote an end which is no part of [their] intention."[10] That end often benefits society overall.

If you don't read Smith carefully, you might think that he's making the same argument as Mandeville and Ivan Boesky—that individual greed is good for society. That's a misreading of Smith, a misreading that was made popular by Ayn Rand. She even wrote a book called *The Virtue of Selfishness*.[11] (At least Mandeville admitted that selfishness was a private vice.) For Rand, greed was the basis for a free economy. Capitalism and greed go together like fat cats and big cigars. Just in case you think she was being cheeky, Rand went out of her way to espouse atheism, and she stridently denounced Christian altruism as antithetical to capitalism: "Capitalism and altruism are incompatible," she said, "they are philosophical opposites; they cannot co-exist in the same man or in the same society."[12] In fact, she had a hard time distinguishing Christian altruism from socialism.

Rand was born in Russia in 1905 as Alisa Zinov'yevna Rosenbaum, and immigrated to the United States in 1925, just as communism was securing its stranglehold on the Soviet Union. Her hatred of the collectivism she saw in her youth was etched into her worldview, her writings, even her strange personality. After coming to the United States, she worked as a scriptwriter in various Hollywood studios. The release of her novel *The Fountainhead* in 1943 made her famous. *Atlas Shrugged,* published in 1957, made her a sensation.

In her novels, she developed characters who expressed her philosophy "of man as a heroic being, with his own happiness as the moral purpose of his life, with productive achievement as his noblest activity, and reason as his only absolute."[13] Her books read more like tracts for her philosophy of "Objectivism" than ordinary novels. As Daniel Flynn puts it: "The themes of Rand's four novels—*We the Living, Anthem, The Fountainhead, Atlas Shrugged*—are identical. As far as the philosophy of her novels goes, to read one is to read them all."[14]

But for millions of readers, the books still work. I discovered Rand during my senior year in college in my final "capstone" course for political-science majors. Rand was assigned reading, sort of. The preapproved reading list was made up entirely of great texts like Plato's *Republic* and Aristotle's *Politics*. But the professor, a rare classical liberal, allowed us students to read one book of our own choosing. So we took suggestions and then voted. Ayn Rand won overwhelmingly; but since we couldn't agree on which of her books to read, we ended up with *For the New Intellectuals,* a collection of excerpts from her most popular works.

Before this class, I had subsisted mostly on the staple of the American college life: derivative left-wing pabulum. It numbed my baloney detector for a while; but after four years, I was growing skeptical. Plato and Aristotle were a welcome relief; Rand was like a blow to the chest. She mercilessly skewered every leftist cliché I had taken for granted.

I found her bracing prose and iconic heroes attractive and repellant at the same time. For a few months, she seized me. I frittered away a week of my senior year reading *Atlas Shrugged* rather than studying for a German final. The book tells about an elite group of creative entrepreneurs and inventors, "individuals of the mind," who go on strike against a state that implements the communist principle "from each according to his ability, to each according to his need." For Rand, these entrepreneurial heroes, like Atlas in Greek mythology, hold up the world. By pursuing their long-term self-interest, they create value for everyone. So when they shrug—that is, strike—society begins to decay.

The hero of *Atlas Shrugged,* John Galt, founds a secret community off the collectivist grid, called Galt's Gulch. Here in this

New Jerusalem, individuality and self-interest are prized above all else. One long chapter of the book, "This Is John Galt Speaking," is nothing but a speech by Galt. It's the perfect distillation of Rand's philosophy. I read it three times.

Despite Rand's official praise of selfishness, however, John Galt doesn't look anything like Ebenezer Scrooge or that fat, cigar-smoking, tuxedo-clad guy in Monopoly. On the contrary, Galt is a pioneer, a brave creator of wealth who pursues his vision despite powerful obstacles, including a malevolent state bent on destroying him. In fact, although Rand despised Christian self-sacrifice, Galt is suspiciously Christlike. He preaches a message of salvation, founds a community, and challenges the status quo and the official powers-that-be, who hunt him down, torture him, but ultimately fail to conquer him.

To be sure, there are dissonant notes. His symbol is not a cross but the dollar sign. The book ends with Galt and his lover tracing the sign of the dollar across a dry valley. But insofar as Galt's character works, it's because he contradicts the miserly stereotype that Rand's philosophy leads the reader to expect. In fact, none of Rand's best fictional characters fit her philosophy very well.

Ayn Rand convinced me that collectivism was a false moral pretense. She also taught me the importance of entrepreneurs in creating wealth. All those supply-and-demand charts and equations and attempts to capture the aggregate behavior with mathematical abstractions that I slogged through in macro- and microeconomics missed the beating heart of capitalism—namely, capitalists. Rand knew that you can't have capitalism without capitalists.

But despite these insights, the Randian spell didn't take. Before long, I realized that she was morally obtuse. She had confused chivalrous self-sacrifice with weakness. It's hard to imagine a bigger mistake. After all, we *admire* the marine who voluntarily falls on a grenade to save his fellow marines, the fireman who charges into a burning building to save people he doesn't know, or the mother who dies protecting her children from an attacker. These are signs not of weakness and vice, but of profound strength and virtue. You don't have to be a Christian to resonate with Jesus's words: "Greater love has no one than this, that someone lays down his life for his friends" (John 15:13). No defense of capitalism that contradicts such an obvious moral truth would work—for me at least.

Only later did I realize that Rand bought into a myth more common among critics of capitalism—that the essence of capitalism is greed.

Myth no. 5: The Greed Myth (believing that the essence of capitalism is greed)

SELFISHNESS AND SELF-INTEREST

Some 30 million copies of Rand's books have been sold, and over five hundred thousand copies of her books are still sold every year. In a poll conducted by the Library of Congress and the Book of the Month Club in the 1990s, *Atlas Shrugged* came in second behind the Bible as the most influential book. Although her work is best known in the United States, it's read around the world. I recently visited Hong Kong, that great monument to free markets and the entrepreneurial spirit, and went into a bookstore to see what people were reading. On the table next to the checkout counter were many of the usual titles you find in bookstores: *An Inconvenient Truth,* by Al Gore, and *The God Delusion,* by Richard Dawkins, were on display. And right beside them, I'm sorry to report, was a Chinese translation of Rand's book *The Virtue of Selfishness.*

Why is Rand so popular? It may be due to a lack of moral defenses of capitalism, and a glut of bad moralizing. Perhaps it's not surprising that many capitalists embrace her: they have nowhere else to go. Who but Rand made industrialists the heroes of novels? It's almost unprecedented. Whatever the reasons for her popularity, however, she completely missed the subtleties of capitalism.

Her hatred of Marxism and collectivism led her to defend a caricature more grotesque than anything Marx imagined.

Her praise of "greed" is the reduction to the absurd of a bad interpretation of Adam Smith's concept of self-interest. Smith, a moral philosopher, didn't goad butchers, brewers, and bakers to be more selfish.[15] He believed that normal adults aren't self-absorbed monads, but have a natural sympathy for their fellow human beings. His point about self-interest is that, in a rightly ordered market economy, you're usually better off appealing to someone's self-love than to their kindness. The butcher is more likely to give you meat if it's a win-win trade, for example, than if you're reduced to begging. Besides, even if you get food by begging, it's degrading. You should avoid it if you can.[16] Smith isn't suggesting that butchers should never help beggars. Unfortunately, most people never read Smith for themselves, but see his words pulled out of context from *The Wealth of Nations* and commandeered for a philosophy he would have rejected.[17]

Smith was a realist. He wasn't naive about the motives of merchants and everyone else. In fact, like most academics, he harbored snobbish prejudices against business. But he knew the difference between self-interest and mere selfishness.[18] Smith believed humans are a mixed breed. We are pulled to and fro by our whims and passions; but we're not slaves to them, since our passions can be checked by the "impartial spectator" of reason. We are capable of vices like greed and of virtues like sympathy.[19]

Moreover, unlike Mandeville, Smith didn't view all our passions as vicious. We may be passionately committed to a just cause, for instance. Still, he saw greed as a vice. So while he agreed with Mandeville that private vices could lead to public goods, he was an ardent critic of the Dutchman: "There is," he said, "another system which seems to take away altogether the distinction between vice and virtue, and of which the tendency is, upon that account, wholly pernicious: I mean the system of Dr. Mandeville."[20] You'd never catch Smith endorsing Gordon Gekko.

For Smith, pursuing your self-interest was not in itself immoral. Every second of the day, you act in your own interest.

Every time you take a breath, wash your hands, eat your fiber, take your vitamins, clock in at work, look both ways before crossing the street, crawl into bed, take a shower, pay your bills, go to the doctor, hunt for bargains, read a book, and pray for God's forgiveness, you're pursuing your self-interest. That's not just OK. In most cases, you *ought* to do these things. Only foggy moral pretense confuses legitimate self-interest with selfishness.

In fact, proper self-interest is the basis for the Golden Rule, which Jesus called the second-greatest commandment, after the command to love God: "In everything do to others as you would have them do to you; for this is the law and the prophets" (Matt. 7:12). I'm supposed to use my rightful concern for myself as a guide in how I treat others. This makes perfect sense, since I know best what I need. "Every man is, no doubt, by nature," Smith said, "first and principally recommended to his own care; and as he is fitter to take care of himself than of any other person, it is fit and right that it should be so."[21]

But self-interest isn't just looking out for number one at everyone else's expense. Since we're social beings, our self-interest includes our friends, families, communities, co-workers, co-religionists, and others.[22] When I pay my bills, I'm not pursuing just my own narrow interest, but the interests of my family, my bank, my community, and whomever I'm paying. I chose my church and my neighborhood and my car not just for myself, but for my children. (Mostly for them, in fact. If I were childless, do you think I'd drive a gray Honda Accord?)

Most of your choices involve the interests of others, too. Just think about it. Self-interest has to do with those things we know, value, and have some control over. I'm most responsible for what *I* do. I have more responsibility for my daughter and next-door neighbor than I have for a random person picked out of the Fargo, North Dakota, phone book. For one thing, I know how to help my daughter. I know nothing of the random person in Fargo. I don't know how my actions affect that person.

If I were equally responsible for everyone, I would have obligations that I have no chance of fulfilling, responsibilities over which I have no control. I could never make a decision in good faith about matters near at hand, since my mind would be swamped by a trillion other facts: a kitten thirsty for milk in Mumbai, a little old lady needing help crossing the street in San Jose, California, a man who has a plumbing leak in Paris, Texas, a bunch of bananas stolen in Guatemala that ends up at a Grand Rapids Meijer store with a Chiquita sticker on it. I don't know any of this. How can I factor them into my daily choices? This doesn't mean I have no responsibility for my fellow human beings in far-off places. It just means that most of my cares and obligations lie closer to home, and extend especially to those matters I know something about.

This is as it should be. If we had to know where all our actions would lead before we could act, we'd do best just to sit and drool. We're accountable to God to act virtuously, and to consider the foreseeable ends of our actions. In a complex market, however, we can't know all the consequences of what we do, any more than we can predict how the proverbial butterfly flapping its wings in Shanghai will affect the weather in Copenhagen. That's just the way it is. Smith's point was not that the more selfish we are, the better a market works. His point, rather, is that in a free market, each of us can pursue ends within our narrow sphere of competence and concern—our "self-interest"—and yet an order will emerge that vastly exceeds anyone's deliberations.[23] The same would be true even if we did everything with godly rather than mixed motives. The central point is not our greed, but the limits to our knowledge. The market is a higher-level order that vastly outstrips the knowledge of any and all of us.

FALLING INTO CAPITALISM

So capitalism doesn't need greed. At the same time, it can channel greed, which is all to the good. We should *want* a social order that channels proper self-interest as well as selfishness into socially desirable outcomes. Any system that requires everyone always to act selflessly is doomed to failure because it's utopian. People aren't like that. That's the problem with socialism: it doesn't fit the human condition. It alienates people from their rightful self-interest and channels selfishness into socially destructive behavior like stealing, hoarding, and getting the government to steal for you.

In contrast, capitalism is fit for real, fallen, limited human beings. "In spite of their natural selfishness and rapacity," Adam Smith wrote, businesspeople "are led by an invisible hand . . . and thus without intending it, without knowing it, advance the interest of the society."[24] Notice that he says "in spite of." His point isn't that the butcher should be selfish, or even that his selfishness is particularly helpful. Rather, his point is that *even if* the butcher is selfish, even if the butcher would love nothing more than to sell you a spoiled chunk of grisly beef in exchange for your worldly goods and leave you homeless, the butcher can't make you buy his meat in a free economy. He has to offer you meat you'll freely buy. The cruel, greedy butcher, in other words, has to look for ways to set up win-win scenarios. Even to satisfy his greed, he has to meet your desires. The market makes this happen. That's making the best of a bad situation, and of a bad butcher.

DOES CAPITALISM MAKE PEOPLE GREEDY?

But even if capitalism doesn't need greed, doesn't it feed greed? According to Jim Wallis, ours is "a culture that extols greed."[25] Like Wallis, many religious scholars don't even distinguish capitalism from greed.[26] Capitalism is just greed elevated to economics, or so they think. And if you happen to catch Donald Trump on *The Apprentice*, you might suspect they're onto something.

To be sure, plenty of capitalism's champions appeal to greed, even glory in it. But there's no evidence that capitalist countries in general, or Americans in particular, are greedier than average. In fact, the truth is just the opposite. A recent British study showed that the United States, which has the fourth-freest economy in the world, is the most generous country when it comes to charitable giving. Americans give about 1.67 percent of our GDP to charities, more than twice the second-ranked United Kingdom (0.73 percent) and Canada (0.72 percent). The French came in dead last in the study, giving just 0.14 percent of their GDP to charity. The study also found an inverse correlation between taxation and giving. The more government confiscates, the less people give. Conversely, the freer the economy, the more people give.[27] This makes you wonder how much Americans would give if the government took less from us.

It's nice when statistics confirm common sense. A poor population is in no mood to be generous. When you're starving, you're desperate. Anyone who visited the Soviet-bloc countries before the collapse of the Soviet Union experienced this firsthand. In such places, only the very virtuous are generous. If you're rich, on the other hand, it's easy to be generous, even if you're morally mediocre. Besides, prosperity gives you something to be generous with. And when the government confiscates your wealth and claims to be using it to help the less fortunate, the situation not only creates resentment; it creates the illusion that you already gave at the office.[28]

Now for the obligatory disclaimer: *of course* Americans should be more generous, more loving, more thankful, more thoughtful, and less sinful. If you look, you can find greed all across the fruited plains and in every human heart. That's because we're fallen human beings, not because we're Americans. Every culture and walk of life has heaping helpings of greedy people. There are greedy doctors, greedy social workers, greedy teachers, politicians, park rangers, and youth pastors. That's why greed can explain why capitalism works no better than it can explain the universal thirst for, say, well-synchronized traffic lights: greed is universal; capitalism is not.

THE GIFT GIVERS

Think of the stereotypical miser like Ebenezer Scrooge (as opposed to the ordinary greedy person). Misers hoard their wealth. They hole up in their bedrooms, counting their gold bullion and hiding it in their mattresses. They clutch it and count it and fondle it, like Smaug in *The Hobbit,* an old dragon sleeping on his pile of booty, knowing when a single gold goblet has gone missing. "Do not store up for yourselves treasures on earth, where moth and rust consume and where thieves break in and steal," Jesus commanded his disciples, "but store up for yourselves treasures in heaven. . . . For where your treasure is, there your heart will be also." Jesus is talking about the person who hoards, who trusts his possessions rather than God. "You cannot serve God and wealth" (Matt. 6:19–24). The apostle Paul said that greed is idolatry (Eph. 5:5). If religion involves our "ultimate concern," as Paul Tillich said, then the miser is an idolater. He worships his money. That's because you can have only one ultimate concern.

Luke records a sermon in which Jesus commands his disciples to stay watchful, to trust God for their needs, to seek first God's kingdom, "and these things will be given to you as well." Jesus wasn't telling his disciples to be scatterbrained and blow off deadlines. He was talking about their—our—ultimate loyalty and trust, which must lie in God and his provision, not with our own plans and possessions (Luke 12). This is a recurring theme in the Bible when it comes to money: we're all in danger of misplaced loyalty, and if we're wealthy, our wealth is a likely recipient of our loyalty. That warning applies to you, and it applies to me.

Let's drill a little deeper. Many of the biblical warnings seem to apply to misers. But how many misers have you met? Do you know anyone who keeps a bag of money in his mattress, where he can count it? Probably not. We see misers on TV; we read about them in children's books and in Dickens. But in capitalist societies, we regard misers as cranks. They hide out in compounds; stockpile food, guns, and gold; see hidden messages in newsprint and black helicopters in the sky. They're in very short supply. That's because capitalism discourages miserliness and encourages its near opposite: enterprise.

"The grasping or hoarding rich man is the antithesis of capitalism, not its epitome," writes George Gilder, "more a feudal figure than a bourgeois one."[29] The miser prefers the certainty and security of his booty. It's right there in his cold hands. Entrepreneurs, in contrast, assume risk. They cast their bread upon the waters of uncertainty, hoping that the bread will return with fish. They delay whatever gratification their wealth might provide now for the hope of future gain. The miser treats his bullion as an end in itself. The entrepreneur, whatever his motives—including the desire for more money—uses money as a tool. The carpenter uses hammer and saw; the doctor, scalpel and stethoscope; the entrepreneur, cash and credit.

Only by the constant din of stereotype could we ever come to mistake the entrepreneur for the miser. In his modern classic, *Wealth and Poverty,* George Gilder compares entrepreneurs not to misers but to tribal chiefs in primitive societies who give gifts and tributes to neighboring tribes in hope of some uncertain future reciprocity. Among the moneyless Siaui in the Solomon Islands, for instance, the *mumi* ("big man") would offer a feast of meat and coconuts, expecting some unspecified return. Imitating this first act of faith, other *mumi* would try to outdo each other by throwing ever larger, even more fitting feasts.

Indian tribes like Alaska's Kwakiutl had a similar ritual of gift giving, called the potlatch. A great gift fits the needs of the recipients, and elicits a response. "In the most successful and catalytic gifts," Gilder observes, "the giver fulfills an unknown, unexpressed or even unconscious need or desire in a surprising way. The recipient is startled and gratified by the inspired and unexpected sympathy of the giver and is eager to repay him. In order to repay him, however, the receiver must come to understand the giver. Thus the context of gifts can lead to an expansion of human sympathies." It's no surprise that such rituals often led to peaceful, voluntary trade among previously warring tribes.

The great missionary David Livingstone described such rituals from his work in Africa. Early in his career, he arrived at a village. The chief informed him that they needed to do a gift exchange. Livingstone had to go first, so he laid out his few belongings, hoping the chief wouldn't take his goat, which provided digestible milk. (Livingstone had digestion problems.) The chief, of course, took the goat. He gave Livingstone a big stick in return. At first Livingstone felt conned. But later he learned that the chief had given him his chiefly staff. It was like being given a signet ring. Now Livingstone could use the carved stick to gain welcome to any of the hundreds of other villages in that region.

Even if such acts of gift giving are not wholly selfless, they rise above the vulgar impulse of envy, which encourages theft and plunder. They also reveal a surprising feature of enterprise: the gift giving *precedes* the voluntary exchange so beloved of economists. After all, before you can exchange, you must have something to exchange. I must have a good or service, a coconut or a potholder or an iPod that someone wants for trade, to ever get started. Right off the bat, if I'm an entrepreneur, I have to think about the wants and needs of others. I must make or acquire something desirable. Great entrepreneurs succeed by anticipating and meeting the desires of others. In that sense, Gilder argues, they are altruistic: *alter* in Latin means "other." (Not for nothing did Ayn Rand dedicate her final lecture to a tirade against Gilder.) Competition between entrepreneurs in a free economy thus becomes an altruistic competition, not because the entrepreneurs have warm fuzzies in their hearts or are unconcerned with personal wealth, but because they seek to meet the desires of others better than their competitors do.[30]

The gift-giving *mumi* were moving beyond the simple certainty of barter, in which one concrete good was exchanged for another. They were taking the first step toward a system of money and investment:

Capitalism consists of providing first and getting later. . . . The mumi became impatient with the tangled negotiations of exchange and started simply donating his product. It worked. He had invented a form of capitalist investment, giving up his wealth in order to save it: part with his goods in order to partake of a growing diversity of goods donated by others. In most cases, the feasts and offerings were essentially entrepreneurial.[31]

The invention of money made it possible to extend faith and trust beyond the natural relationships of kin and tribe.[32] In fact, some measure of trust must exist before you can have money.

Think about it. Money can be exchanged for a good or service only if a community accepts it as payment for a debt. Without the trust that others will accept the same money you accept, that money becomes mere metal or paper. So the acceptance of money between individuals is a sign of a preexisting trust. Money also confers more freedom than barter, since you don't have to spend it on anything in particular.

Money makes entrepreneurial investments possible, since it allows the entrepreneur to compare the value of different opportunities using a standard unit of measurement, just as a square foot allows you to compare the size of two plots of land, even if they're on opposite sides of the planet.

While money represents bonds of trust within a community, an investment is an act of trust that extends into the unknown future. Like the coconuts of the *mumi,* investments are a gift. Such gifts don't conform to a selfless ideal, of course, but rather to what anthropologist Claude Levi-Strauss called the Law of Reciprocity: "The essence of giving is not the absence of expectation of return, but the lack of a predetermined return." Like gifts, capitalist investments are made without a predetermined return.[33]

Many Christian critics of capitalism look at capitalists and see only the superficialities. They notice that entrepreneurs work with money and seek to multiply it. They associate greed with money, money with entrepreneurs, and so greed with entrepreneurs. This simpleminded reasoning fails to distinguish the modern entrepreneur from the medieval miser. Entrepreneurs don't look *at* their money but *through* it to what it can do. The money in hand is a means, not an end, even if one of the possible ends is more money. They risk actual wealth in the hope of multiplying it. They pursue visions; they seek to create something they imagine may fulfill some need or desire—a stirrup that makes it easier to ride horses; a new way of dispersing risk between partners in international shipping; a nut that tastes good when it's toasted, mashed, and spread out on bread; an intuitive user interface for a computer with little "windows" and other images that replaces obscure code instructions; adhesive that's sticky enough to post a note but not too sticky, so you can remove it when you're done; cat litter that makes it easy to keep your cat indoors. Such creations half-form in the mind of the entrepreneur long before they show up on the shelf at Safeway.

Take what was surely the least glamorous invention in the previous list. In 1940, 50 million Americans weren't pining away for cat litter. No one had even thought of it. Most people kept their cats outside. (The few indoor cats had to use sandboxes or messy ashes to relieve themselves.) Kitty Litter is actually a brand name of a product first "invented" in 1947 by a man named Edward Lowe. In retrospect, it's obvious that Lowe was a born entrepreneur. But his story started with the plainest of circumstances. After leaving the navy at the end of World War II, Lowe got a lowly forty-five-dollar-a-week job working for his father, delivering sand, gravel, sawdust, and clay throughout southern Michigan. In 1946, his father's company started selling absorbent clay called fuller's earth. Ed Lowe tried to market it as nesting material for chickens under the name Chicken Litter. The idea bombed with farmers, though Lowe always maintained that the chickens really liked the stuff.

It seemed like a failed idea, but during the frigid Michigan winter of 1947, a friend asked Lowe for something that would allow her cat to relieve himself indoors. Lowe realized that the sand was frozen and ashes were messy, so he gave her a five-pound bag of Chicken Litter. She used it and then came back for more. A few months later, he decided to pitch the product to the owner of the Davenport Pet Shop, who dismissed it as "dirt in a bag." Undaunted, Lowe left ten bags with the owner and told him to give it away. Three weeks later, the owner called to ask for more.

You can guess the rest of the story, at least in broad outline. Lowe was a tireless marketer, so practically every pet store in the United States eventually sold Kitty Litter. Lowe later marketed the product as Tidy Cat for the new supermarket pet aisles while keeping the Kitty Litter brand for pet stores. Other companies elaborated on and tried to compete with Lowe's original concept. The most successful is clumping litter, invented by biochemist Thomas Nelson in 1984. Cat litter is now a billion-dollar industry that transformed the way in which we keep cats as pets, and no doubt contributes to the fact that there are now more cat owners than dog owners in the United States.[34]

If Lowe had proposed his Kitty Litter to the celebrity judges on *American Inventor,* he probably would have gotten four no votes. No one, not even Lowe initially, imagined the nature and extent of the market for cat litter. But Lowe *anticipated* and then cultivated that market. Here as elsewhere in enterprise, initiative and invention preceded supply. Supply preceded demand. Indeed, it created demand. Lowe didn't look at a market and figure out a way to fill it. He imagined a market for a product and then *created* the product and, in a sense, the market, too. The supply-and-demand charts that fill economics textbooks and haunt the memories of former econ majors can't explain anything until there is a supply. At the base of the capitalist system is not greed or consumption, but intuition, imagination, and creation. As Thomas Edison said, "I find out what the world needs . . . then I proceed to invent it."

In fact, entrepreneurial capitalism requires a whole host of virtues. Before entrepreneurs can invest capital, for instance, they must accumulate it. So unlike gluttons and hedonists, entrepreneurs set aside rather than consume much of their wealth. Unlike misers and cowards, however, they risk rather than hoard what they have saved, providing stability for those employed by their endeavors. Unlike skeptics, they have faith in their neighbors, their partners, their society, their employees, "in the compensatory logic of the cosmos."[35] Unlike the self-absorbed, they anticipate the needs of others, even needs that no one else may have imagined. Unlike the impetuous, they make disciplined choices. Unlike the automaton, they freely discover new ways of creating and combining resources to meet the needs of others.[36] This cluster of virtues, not the vice of greed, is the essence of what the Reverend Robert Sirico calls the "entrepreneurial vocation."[37]

Very few critics of capitalism understand this. But the problem isn't just with the critics. Too often, even the fans of capitalism neglect the entrepreneur, focusing on free markets rather than free men and women.[38] Adam Smith, for instance, was rightly intrigued by the invisible hand. Most economists have followed Smith, preferring to study the measurable market rather than mysterious man. Too often, elusive entrepreneurs get replaced with simple abstractions like supply and demand. It's easy to see why this picture is so attractive. If economists can assume that everyone is producing and trading according to something simple like "self-interest," they have some hope of predicting what will happen in an economy. Everything is accounted for, and we don't have to worry about the virtues, vices, and surprises of individual businesspeople. In short, we get capitalism without capitalists.

Unfortunately, this desire to conform economics to a narrow definition of science has discouraged most economists from focusing on the creative agents that give the economy life. And not just any agents: *entrepreneurs*. In doing so, however, economists ignore the greatest source of wealth creation in capitalism. As George Gilder says: "Man . . . and not mechanism is at the heart of capitalist growth."[39] And man is not a mechanism. Unlike the orbit of the planets, which astronomers can predict with uncanny accuracy, what an entrepreneur will create is unpredictable ahead of time—even to the entrepreneur himself! The process is inherently unpredictable. We see the work of entrepreneurs after the fact. Supply and demand kicks in after there's a supply. And in a modern economy, supply depends on enterprise. The predictive methods of science are not up to the task of capturing enterprise in a petri dish or a telescope or an equation. So the entrepreneur is a source of trouble for any economist who wants economics to be a completely predictive science.

Even free-market economists often forget this point. They're right to defend freedom. Enterprise requires a free society. After all, if entrepreneurs aren't free to save, gather, and risk capital in pursuit of their visions, they can't create new wealth and opportunity. Without freedom, there can be no creativity.

Free-market economists are also right to glory in the wonders of the market's order, which distributes goods, services, and information better than any so-called planned economy. For all its wonders, however, the invisible hand of the market would not emerge without the visible hands of the entrepreneurs who first create new value. As George Gilder has said, "Without entrepreneurs, economies are dead."[40] "I, Pencil" and my own "I, iPod" are great for illustrating the organizing power of the free market, but they presuppose the true sources of wealth creation. Sure, no one person can make a pencil or an iPod. They come at the end of a process so complex that no one could coordinate it. Nevertheless, neither pencils nor iPods would exist if some*one* had not first had the idea of a pencil or an iPod.

Actually, that's not quite right. What came first was not the idea of a pencil or an iPod, but a pencil-like or an iPod-like idea. If you study the stories of great inventors, you discover that the actual inventions are usually different from the initial idea. How could it be otherwise, since we can't peer into the future? All we can do is anticipate it. That's what Ed Lowe did on the way to Kitty Litter. Nevertheless, the entrepreneur's serendipitous shot in the dark is the first irreplaceable step. By itself, of course, a mere idea is worthless. Great inventions, by themselves, accomplish very little. To create new wealth, a good or service must be put to use. So somebody must implement and market the idea. That's the job of entrepreneurs.

Most entrepreneurs seek new and improved ways to meet a need or desire that is already being met. They seek to design a car engine that gets more miles to the gallon, to offer a life-insurance package with lower rates and better benefits, to build a better mousetrap. These entrepreneurs tweak around the edges by looking for ways to compete in a market that already exists.

But some entrepreneurs make vast leaps of daring and imagination, *creating* the markets and even the needs that they then satisfy. The process can be so mysterious that even the key players don't quite realize what they're doing. They never imagine all the future applications of their work. Consider e-mail. Twenty years ago, I didn't need e-mail. Almost no one did. We needed to be able to communicate, of course, but we used telephones and the ordinary postal system for that. But a series of inventions and insights by many people quickly gave rise to e-mail, which no one person invented. It quickly changed the world.

It started in 1965 when different users of the same mainframe computer needed to communicate with each other. This expanded to networked computers and the ARPANET (Advanced Research Projects Agency Networks), so that what we now call e-mail software evolved along with what became the Internet. It was Ray Tomlinson, sometimes mistakenly called the inventor of e-mail, who thought of using the at sign (@) to distinguish users from their computers. That was 1971. In 1993, America Online and Delphi made e-mail available to their subscribers. Then, suddenly, within a few years, that tangle of technologies reorganized and expanded the ways in which human beings interact globally. Now hundreds of millions of jobs, including my own, depend on easy access to e-mail. And since I need my job, that means I now need e-mail—if not to survive, then at least to do my job.

Without entrepreneurs, very little of what we take for granted in our economy and our everyday lives would exist. Here in my office, the concrete forms of entrepreneurial imagination are everywhere: paper, scissors, pens, highlighters, ink, CDs, an empty Tupperware container that held the pork loin I ate for lunch, a flat-screen monitor, fonts, lamps, lightbulbs, windows, drywall, speakers, a laptop computer, a wall painted in Benjamin Moore flat Chestertown Buff. Behind all these visible objects lie real but less visible innovations in finance, manufacturing, and transport that I scarcely comprehend. All of these things are gifts of entrepreneurs. The only one I know about is Ray Tomlinson—because I looked him up. I have no idea where this other stuff came from. Only the most miserly moralizer could look at this mysterious efflorescence of cooperation, competition, and creativity—of entrepreneurial capitalism—and see only the dead hand of greed.

Six

Hasn't Christianity Always Opposed Capitalism?

I still remember my first experience with banking. I was in fifth grade and had saved up $100 from collecting soda bottles (which fetched five cents apiece). My dad took me out of school to have lunch and start a savings account at First National Bank of Amarillo. (Getting out of school probably etched the experience in my memory.) I learned that by depositing the $100 in a savings account paying 6 percent interest, and then letting it sit, I would have $106 by the end of the year. If I let the whole $106 sit for another year, my stash would then be worth $112.36 rather than just $112. That's because of compounded interest. In the second year, I would be gaining interest not just on the original $100, but on $106. If I let that sit another year, the stash would be worth $119.10, and so on until I was millionaire. Or until I needed a new bike—which, of course, came much sooner.

Later, I learned that my dad had simplified the story a bit. My savings was compounding more frequently than once a year. That meant the original money, and the interest on the interest, grew even more quickly than I realized. I knew nothing else about banking, so it seemed like magic: you put some money in, let it sit, get some more, and then get more of the more. Even at a modest 6 percent interest, a kid who invested $100 at the age of ten could have $3,200 sixty years later, just from that initial $100 investment.

Since the arrangement worked out in my favor, I didn't ask skeptical questions. But it's really strange when you think about it. If the same kid took up lawn mowing, socked $4,000 away in a corporate bond fund, and averaged 7.2 percent a year over a period of sixty years, he could have $256,000 at the end of those sixty years. If he managed 10 percent interest a year, at the end of fifty-eight years he'd have a million dollars. Something weird is going on here.

Even Einstein once said that compound interest is one of the greatest mysteries in the universe. And yet, like the water a fish swims in, we rarely notice it or the other effects of modern finance all around us. We put our money in banks, charge our groceries on credit or debit cards, write checks, use PayPal to buy stuff online, and take out fixed-rate mortgages on our home. We buy stocks and bonds, and set aside part of our paychecks for IRAs, 401(k)s, and pension plans. Each of these acts involves, in one way or another, charging or receiving interest on money.

Few of us lie awake at night worrying that we're going to hell because of it. Yes, we know that it's bad to acquire too much consumer debt. You don't want to pay Visa 20 percent interest just so you can get the newest PlayStation, Jet Ski, or half-price liposuction. But buying and selling money *by itself* doesn't trouble most of us. And yet for centuries all the greatest Christian theologians and philosophers believed that charging interest on money was an egregious sin—called usury.

WHAT'S WRONG WITH SELLING MONEY?

Although some groups of Christians have exalted poverty as an end in itself, overall, Christianity has never had a problem with business investments or even with making a profit. For a long time, though, Christianity did object to charging interest on money—that is, to making a profit on a money loan. According to the most learned churchmen, if you tried to "sell" money, you were committing usury. This was the unanimous view of the Christian world throughout the Middle Ages. Dante, in his *Divine Comedy* (written between 1308 and 1321), put usurers in the seventh ring of hell along with blasphemers and sodomites! He saw usury as a serious sin—a type of fraud motivated by greed.[1] But why?

Well, first of all, the Bible seems to say so. In Exodus, shortly after God delivers the Hebrews from slavery in Egypt, he gives

Moses a list of commands to deliver to his people. The Ten Commandments lead the list, but a series of many other "ordinances" follow in their train. They prescribe the death penalty for serious crimes like murder, kidnapping, and bestiality and require just compensation for lesser offenses like letting your ox graze on someone else's land. They also prohibit charging interest on money.

Following the command not to oppress resident aliens ("for you were aliens in the land of Egypt"), God says: "If you lend money to my people, to the poor among you, you shall not deal with them as a creditor; you shall not exact interest from them" (Exod. 22:25). It's the same story in Leviticus, when God is describing the rules that should apply when he brings the Hebrews into the Promised Land:

> If any of your kind fall into difficulty and become dependent on you, you shall support them; they shall live with you as though resident aliens. Do not take interest in advance or otherwise make a profit from them, but fear your God; let them live with you. You shall not lend them your money at interest taken in advance, or provide them food at a profit. (Lev. 25:35–37)

Of course, these commands apply to Jews lending to other Jews. Deuteronomy allows Jews to charge interest to foreigners:

> You shall not charge interest on loans to another Israelite, interest on money, interest on provisions, interest on anything that is lent. On loans to a foreigner you may charge interest, but on loans to another Israelite you may not charge interest, so that the LORD your god may bless you in all your undertakings in the land that you are about to enter and possess. (Deut. 23:19–20)

But the Old Testament still treats the whole business as unsavory. Pithy Psalm 15 asks: "O LORD, who may abide in your tent? / Who may dwell on your holy hill?" The answer:

Those who walk blamelessly, and do what is right,
And speak the truth with their tongue . . .
Who do not lend money at interest,
and do not take a bribe against the innocent. (Ps. 15:1, 5)

Though the New Testament says little about charging interest, Jesus admonishes believers: "If you lend to those from whom you hope to receive, what credit is that to you? Even sinners lend to sinners to receive as much again. But love your enemies, do good, and lend, expecting nothing in return" (Luke 6:34–35). Not exactly something you'd inscribe in granite over the entrance to the Bank of America building.

In addition, the Bible and especially the Gospels are chock-full of warnings about the dangers of money. At one point, Jesus even drove money changers out of the temple with a whip of cords, along with "people selling cattle, sheep, and doves" (John 2:13–16). And Paul warned Timothy: "The love of money is the root of all evil" (1 Tim. 6:10).

Christian scholars in the early and medieval church quoted these Bible passages in almost every discussion about usury. But they didn't just read the Bible. They were also immersed in Roman and Greek thought, which looked askance at charging interest for money. All the classical bigwigs—including Greeks like Plato and Aristotle (who influenced Christian theology in the Middle Ages) and Romans like Cato, Cicero, and Seneca—condemned usury.

The prohibition on usury wasn't limited to the Western world, either. Buddha condemned it in the East, as did Muhammad in the Middle East. In fact, Islam still follows Muhammad in condemning interest. So even though interest charging and primitive banking occurred, a pall of vice still hung over the practice. We shouldn't be surprised that Christian scholars, too, thought that charging interest on money was immoral.

While the Scholastics took biblical prohibitions for granted, they sought to defend God's revealed law with rational argument. No Scholastic worth his salt would settle for "The Bible says it. I believe it. That settles it" unless he couldn't think of any arguments. They trusted God's wisdom. As a result, they asked the next logical question: "Why did God command this? What was the good reason for the prohibition?" And Scholastics could always think of some explanation, even if it wasn't a very good one. All the Scholastics agreed that charging interest was wrong, but they disagreed on why it was wrong.

Early on, Pope Leo the Great (who served from 440 to 461) forbade clerics from taking usury and declared that laymen who did so were seeking shameful gain. This reflected the general view at the time, which was that usury was uncharitable or greedy. As trade and commerce grew, however, so did scholarly thinking on usury. Twists and turns abound over the centuries, but by 1187, the basic contours of the medieval view of usury were in place:

> (1) Usury is whatever is demanded in return in a loan beyond the loaned good itself; (2) the taking of usury is a sin prohibited by the Old and New Testaments; (3) the very hope of a return beyond the good itself is sinful; (4) usuries must be restored in full to their true owner; (5) higher prices for credit sales are implicit usury.[2]

I confess that I found all this business bewildering, so I decided to read John T. Noonan's definitive study, *The Scholastic Analysis of Usury,* published by Harvard University Press in 1957. It's not for the faint of heart. Nuances and hairsplitting fill the Scholastic literature on usury. Subtleties often hinge on the difference between Latin terms like *mutuum, census,* and *lucrum cessans.* But the book convinced me that the ban on usury wasn't a thoughtless residue of a worn-out tradition or the uncritical application of a few Bible passages taken out of context.

BEHIND THE SCENES

To understand the long debate over usury, we have to keep in mind the historical context and the unstated beliefs. Economically, ancient and medieval Europeans were not all that different from ancient Israelites. They weren't trapped in the Stone Age: both used money rather than bartering. But most people still farmed the land—indeed, subsisted on the land, sometimes just barely. Some people fished, worked at crafts like carpentry, and traded basic goods in open markets in cities, but most people lived on land outside city walls. Almost everyone lived and traded only within their extended families and tribes. So the way they interacted was more informal and reciprocal, as befits a family, and less shaped by market forces.[3] Economic growth, such as there was, crept along so slowly you could hardly see it.

By modern standards, almost everyone was dirt-poor. Only the rich, a tiny minority, had any money to lend. Any money lending, then, would involve rich people lending to their poor neighbors, probably their kin, for a basic need like food.

The early Christian world, like the Roman world before it, tended to see money as sterile, functioning only as a means of exchange and without value in itself. And at the time, it largely was. People hid extra money. So while a person might be entitled to have his money returned to him, it seemed uncharitable to charge a poor person for temporarily using money that would otherwise just be collecting dust. After all, money doesn't really wear out like clothing or a house. If somebody wears your clothes for a year, you can't get your original clothes back. So you can rightly charge rent for your clothes. Since money is a unit of exchange, however, an *amount* of money can be repaid exactly (even if the debt is repaid with different coins). Charging for the monetary units would be like charging for inches or minutes. And charging huge interest rates that couldn't be repaid would add insult to injury, since it would exploit a person's bad fortune and ignorance. Thus, given the historical context and the belief that money was sterile, the ban on usury made a lot of sense.

Around the twelfth century, however, trade began expanding between cities and territories throughout Europe, leading to a greater division of labor and higher productivity. This created several problems: First, growth in trade will lead to a shortage of gold and silver coins—the common form of currency. After all, there's just so much of the stuff to go around. Second, it's hard to make large exchanges over hundreds or thousands of miles when money is in the form of heavy gold and silver coins that can be easily stolen or lost in a shipwreck. Finally, the different coins used in Bruges, Milan, and Rome were often reminted and debased with less-valuable metals, so the ordinary person could easily get ripped off by unscrupulous merchants or kings.

Out of these necessities, the bank as we know it was invented.[4] The despised job of the money changer was crucial here, since money changers knew how to compare florins, guilders, and pounds, and to separate them from the fakes. Money changers eventually began keeping deposits for various clients, so that when two clients made an exchange, all the money changer had to do was credit one account and subtract from the other. Simple arithmetic had replaced a risky and cumbersome movement of coins.

Eventually banks emerged with branches in different cities. This gave merchants a way to transfer payment safely over large distances, since banknotes stood in for the money stored safely in a bank vault. Sophisticated banks of this kind first appeared in the city-states of northern Italy and spread from there to Flanders (modern-day Belgium and the Netherlands) and England. By the fourteenth century, there were some 173 major banks in cities such as Florence, Pisa, Genoa, Lucca, Venice, and Milan.[5]

Sociologist Rodney Stark explains the next step:

Local bankers began to credit and debit from the accounts of one another's depositors—as with modern checking accounts. . . . When such a transfer was made over a considerable distance, it involved a *bill of exchange*—a notarized document authorizing payment to a specific individual or firm. To settle payment for wool cloth shipped from Bruges [in Belgium] to Genoa [in Italy], for example, a bill of exchange was sent to a bank in Bruges from a bank in Genoa, whereupon the Bruges bank credited the account of the woolen firm and entered this in its books as a credit held against the Genoese bank. Being merely a sheet of paper and of no value except to the bank in Bruges, the bill of exchange could be rapidly and safely transported.[6]

The process became so common that not only merchants, but governments and even the pope, used banknotes to pay bills. In fact, some banks had such large deposits that they could lend money to kings. What was fit for a king was soon fit for the commoner. Individuals and firms with extra money began entrusting banks with their money, which they would withdraw as needed. This could happen only once people were convinced that their money was safer in a bank than hidden in a mattress or a hole in the ground. So banking grew only as bonds of trust grew beyond family and ethnic lines to connect larger and larger groups of people.

Eventually, banks realized that, if they had enough depositors, they didn't need to keep all their deposits on hand to meet the day-to-day demand from depositors. They could lend some of it out, not literally by handing out coins, but by circulating more banknotes than they held in reserves as coins and precious metals. They were issuing credit, and over time, all sorts of credit forms developed, for purchasing land, equipment, and other forms of capital.

If you're following closely, you'll notice that somewhere along the way, the world changed, at least in Europe. Society went from poorer, smaller, somewhat isolated kin-based communities to larger, more diversified trade economies, where more people and places could focus on their "competitive advantages," on doing what they were the most competitive at doing. Money started as a store of value and a unit of exchange to overcome the coarseness of barter. Trade and division of labor increased the total amount of wealth, leading to a surplus of money far beyond the hoards of thieves and tyrants.

Banks then started as a place to hold money securely and to allow individuals, organizations, and governments to make accurate and safe exchanges over large distances. Popes, kings, and wool merchants could pay their bills at a distance without worrying about thieves in the high mountain passes stealing their bags of gold. Banks did this by *representing* money with math and paper. Over time, more and more people trusted banks to serve these functions, so they deposited their money in banks rather than hoarding it. Banks began using their cash reserves to issue credit. In doing this, they were basically creating more money (not just printing bills, which is a different thing). That credit was used to start and support new enterprises. The surplus money of a few was no longer hoarded and unproductive, but set free for others to use creatively. Banks were using accumulated wealth to create more wealth by functioning as brokers between depositors and investors.[7] They were making "more of the goods of the world available to all."[8] Without anyone really intending it, money had been transformed—sublimated. It was clearly not sterile, but *fertile*.

In the 1300s, insurance also was invented. This allowed investors who used ships to transport their goods to spread out the risk of a shipwreck. Everyone paid a little bit up front, just in case disaster struck. If the ship with its cargo went down, the insurer would pay out the benefit. The sharing of risk prevented any one investor from suffering catastrophic loss. This, in turn, attracted more investors into the business of international trade. Between banking and insurance, the buying and selling of risk allowed many new ventures and enterprises to be created.

Of course, banks and insurers charged for their services. They were assuming risk, and so, quite rightly, they needed to charge something to offset the risk. Banks also paid interest to depositors, who assumed some risk by depositing their money in a lending bank. The complicated math needed to keep track of interest encouraged bookkeepers (and eventually everyone) to switch from cumbersome Roman numerals to the infinitely versatile Hindu-Arabic numerals that we now use. Banks had become far more than places to store money. Banking had become a highly symbolic enterprise. This might have seemed less obvious when most money was a commodity like gold or silver. It's easier to see today when a single small piece of paper, whose ingredients are worth next to nothing, can nevertheless represent ten thousand dollars. With such strange objects floating around our economy, it should be crystal clear: over time, finance and banking become more and more representational—and immaterial.

But these ethereal signs and numbers shape the real world. In the Middle Ages, the presence of banks transformed even ordinary loans between individuals. Money came to have what economists call a "time value." As long as it outpaces inflation, money in an interest-bearing account becomes more valuable over time. So if I lend my neighbor money that I could put in a savings account or a 401(k), bearing interest, I'm not just risking the money. I'm forgoing other opportunities to make more money.

Notice what happened: In many cases, charging interest for money was no longer a rich man lending his poor kinsman a few unproductive coins to buy food. There were *capital loans* available from banks. Unlike the invention of the car, the lightbulb, and the cotton gin, the event got no press. In fact, it took centuries for society to fully realize its significance. Many of us still don't get it.

At some point, though, the old ban on usury started to stick out like a sore thumb. It slowly dawned on people that money lent for capital was different from money lent to a poor neighbor out of need. When banks charge interest on a loan or insurers charge for coverage, they're charging for something. By lending the money, for instance, the bank is forgoing other opportunities to use the money, and it's taking a risk in lending the money in the first place.

In their careful reflections on usury, the Scholastics had anticipated some of these developments. But it still took centuries of disputes for scholars to work out the details and to clearly distinguish ordinary banking from usury. After the Reformation, some Reformers, such as John Calvin, were quick to modify the ancient ban on interest.[9] And Catholic scholars eventually did so as well.

The church didn't decide that usury was OK, however. Rather, it became much more precise in defining usury. Usury isn't charging interest on a loan to offset the risk of the loan and the cost of forgoing other uses for the money; it's unjustly charging someone for a loan by exploiting them when they're in dire straits.[10] That's the work of loan sharks, not banks.

Most Christians now distinguish loan-sharking from ordinary banking. A few still don't; they believe the usury myth.

Myth no. 6: The Usury Myth (believing that working with money is inherently immoral or that charging interest on money is always exploitive)

This myth is much less common than the others we've discussed. Still, it shows up behind the scenes. Just think of the way stockbrokers and bankers are portrayed. When I think of a generic banker, I still picture the skinflint bankers in *Mary Poppins,* with the stiff white collars and black suits, and the bitter Mr. Potter in *It's a Wonderful Life.* No one complains about such stereotypes. And such stereotypes aren't confined to the media. They're just as common from religious leaders. A feminist Web site (Sunshine for Women) recently featured a sermon called "What Would Jesus Do." It started with a list of things Jesus would *not* do. We're told that Jesus would not wear Gucci or own a Fortune 500 company. We're also told:

> Jesus Would **NOT** be a Wall Street trader, a banker for a large national or international banking conglomerate, or participate in the World Bank or the International Monetary Fund (IMF).[11]

If this doesn't strike you as strange, substitute "doctor" or "tailor" or "candlestick maker." Officially, the usury myth may have died, but it lingers on as the vague prejudice against people who work with money.

THE BIRTH OF CAPITALISM

Although few people now view usury as Christian scholars did in the Middle Ages, many people—believers and unbelievers, critics and fans of capitalism—think the long ban on interest shows that Christianity was opposed to capitalism and hindered it from emerging. As a result, they think capitalism is the creation of the secular modern age.

There's just one problem with this historical gloss: it's wrong. For a thousand years, practically everyone everywhere thought charging interest on money was immoral. And bans on interest still hold sway in some parts of the world. What's interesting about the Christian West is not that it once condemned all charging of interest, but that it eventually learned to make careful distinctions and develop vibrant, wealth-creating capitalist economies with sophisticated banking systems. John T. Noonan, in his distinguished study of usury, argues that, despite appearances, the usury debate actually gave rise to modern banking and economics by giving the West unique insights into the nature of money and commerce: "The scholastic theory of usury is an embryonic theory of economics. Indeed it is the first attempt at a science of economics known to the West."[12] So if anything, the tedious debate over usury gave the West a head start in developing capitalist economies.

In fact, many have argued that Christianity helped give birth to capitalism. The most famous argument along these lines is Max Weber's book *The Protestant Ethic and the Spirit of Capitalism*.[13] Weber, a German sociologist, argued that capitalism grew out of the Puritan strain of Calvinism. (We get the phrase "Protestant work ethic" from Weber.)

His basic argument goes like this: Calvinism teaches predestination, which means that if you're a member of the elect, you can't lose your salvation. You're saved not through any work on your part, but because God has decided that you'll get in. So you'd think Calvinists would rest secure in their salvation, unlike Catholics, who can say only that they hope they're saved. But the doctrine of predestination doesn't tell you if *you're* one of the elect, since Calvinism teaches double predestination: the elect are predestined to salvation, the reprobate to damnation. Worse, according to Calvin, some people think they're elect when they're really reprobate. Bummer.

Weber argued that this dilemma caused Calvinists to seek visible signs to verify that they were, in fact, elect. Calvinists, especially Puritans, came to see economic success as just such a sign. Much more than other Christians, they affirmed worldly professions as every bit as much a part of God's calling as the pastoral ministry. Just as God might call a missionary to go to Papua New Guinea, he might call an entrepreneur to build textile mills. Both the missionary and the entrepreneur, in the Calvinist view, were pursuing God's sacred calling for their lives.

This belief made Calvinists enthusiastic and successful businesspeople. But it didn't create conspicuous displays of wealth, since Calvinism encouraged frugality and investment rather than instant gratification and luxury. Calvinists tended to reinvest their wealth rather than sinking it into grandiose palaces and fine banquets. According to Weber, this paradoxical mix created the unique "spirit of capitalism."

Weber is important for several reasons. He resisted the intellectual fashions at the time, which tended to ignore the influence of ideas, especially religious ideas, on culture and society. He was discerning enough not to mistake capitalism for simple greed. Greed is universal, but capitalism is not. He understood that mere greed leads to theft or instant gratification, not capitalism. A thriving capitalist economy needs entrepreneurs who will save, risk, invest, and hope in the future, not clutch their hoard like Ebenezer Scrooge. And he was certainly correct that capitalism and Protestantism have often prospered in the same places, especially in the English-speaking world.

Nevertheless, scholars were quick to criticize his argument. Religious scholars pointed out that his understanding of Puritanism was idiosyncratic, and seemed to reflect his experience with his religious father rather than with any careful study of Calvin or Calvinism. In fact, his book contains precious few references to back up his argument.

But the most obvious problem with Weber's thesis is historical: capitalism can't come exclusively from Calvinism for the simple reason that thriving capitalism appeared in Europe centuries before Calvin. Remember banks. A stable banking system encourages thrift and savings over instant gratification and consumption—one of the basic requirements for capitalism. But banks didn't spring up out of nowhere in Calvin's Geneva. They showed up some two hundred years earlier in the Catholic city-states of northern Italy.

Weber should have located the sources of capitalism not in Calvinism alone, but in Christianity more broadly. A contemporary sociologist, Rodney Stark, has done just that. In *The Victory of Reason: How Christianity Led to Freedom, Capitalism, and Western Success,* Stark argues (like Weber, against current fashion) that a society's beliefs, far more than its real estate, shape its destiny. That claim alone is controversial; but Stark doesn't stop there. He bucks a second trend, arguing that we in the West owe our political and economic freedoms to Christian ideas: "The success of the West," Stark argues, "including the rise of science, rested entirely on religious foundations, and the people who brought it about were devout Christians."[14]

This contradicts the received wisdom, which considers Christianity the benighted enemy of freedom and progress. Even the words we use to describe various historical periods betray an anti-Christian stereotype. There were the Dark Ages, after enlightened Rome collapsed and the church darkened the minds and imaginations of Europe. Then came the Middle Ages, when things got a little better but were mainly just a way station on the path to some other age that really mattered; and finally, we reached the Renaissance (from an Old French word meaning "rebirth") and the Enlightenment, when sweet reason supposedly broke free from the shackles of faith to give us human rights, freedom, and prosperity.

Weber partially resisted this stereotype, but only partially. Weber's thesis was, in a sense, just an interesting twist on the prevailing opinion, since his argument was strongly anti-Catholic.[15] Though officially the cartoonish, anti-Christian version of how the West arose has been out of favor with professional historians for some time, it still clouds our perceptions and our language, much as Freudianism does. Even skeptics of Freud, for instance, still find it difficult to avoid referring to the superego, the sub-conscious, and oral fixations. Similarly, historians still find it hard to describe the West without blaming Christianity for everything from slavery to monarchs.

Stark does not just tweak but flatly challenges this official story in almost all of its particulars. He focuses not on Christianity generically, but on the way in which Christianity nurtured faith in reason and progress. "The rise of the West," Stark argues, "was based on four primary victories of reason":

1. Faith in progress within Christian theology

2. The way that faith in progress translated into technical and organizational innovations, many of them fostered by monastic estates

3. The way reason informed both political philosophy and practice to the extent that responsive states, sustaining a substantial degree of personal freedom, appeared in medieval Europe

4. The application of reason to commerce, resulting in the development of capitalism within the safe havens provided by responsive states[16]

Stark especially identifies the development of systematic theology, "formal reasoning about God," in Christianity. He argues that such intellectual exercises were not trivial, but eventually led to tangible social progress.

To defend his thesis Stark spends much of his time describing the profound cultural and technological innovations that emerged in the so-called Dark Ages. Despite centuries of bad climate, during this time improved water mills, windmills, horse collars and horseshoes, wheeled plows, chimneys, eyeglasses, clocks, stirrups, the magnetic compass, and many other inventions sprang up. Similarly, education and capitalism emerged not with the Reformation or the Enlightenment, but in medieval monasteries. And, as we've already seen, finance and banking emerged in northern Italy's city-states centuries before Luther nailed his ninety-five theses to the Wittenberg door.

Woolen cloth first brought capitalism to northern Europe; and capitalism continued to prosper there after it was repressed by despots in southern Europe. At this point, Protestantism enters the story. Since much of the north became Protestant, however, it was easy for historians to associate capitalism with Protestantism, and anticapitalism with Catholicism. Stark explains why the colonies of the New World had such different fates. It wasn't merely that Spain was Catholic and England was Protestant. The significant difference, he argues, was between the Spanish and British colonial outlook. Spain was given to despotism, while Britain was, by comparison, much more liberal. Therefore, these powers built profoundly different empires. "The British colonies were founded on production," Stark writes, "the Spanish colonies on extraction." Such extraction propped up the Spanish dynasty for a while, but it failed to create wealth. So, not surprisingly, it eventually failed, and it left the Spanish colonies in political and economic disarray. The former British colonies, in contrast, have largely succeeded.

Stark probably focuses too much on reason. Other Judeo-Christian themes clearly contributed to capitalism as well. Here are just a few:

- The idea that God's creation is good, even if marred by sin

- The idea that private property is right and proper, not a material evil

- The Christian moral code as a whole

- An optimism about the future tempered by the doctrine of original sin—which together encouraged hard work, investment, innovation, limited government, checks and balances, and a distrust of utopian schemes

And yet, despite Stark's focusing too narrowly on reason, his wider point seems right: contrary to stereotype, Christianity provided the prerequisites for a vibrant capitalism.

Armed with solid statistics, Stark also argues that nations that protect property rights, individual freedom, and freedom of religion—like the United States—actually *encourage* religiosity more than do those countries—like much of Europe—that have state churches and less regard for private property and individual rights.[17] So much for the claim that capitalism inevitably gives rise to secularism. Secularism is a problem in the United States, of course, but a free economy is not its cause.

WHAT DOES THE BIBLE REALLY SAY ABOUT USURY?

So we have good historical reasons for seeing Christianity as an important source for capitalism. But we're not out of the woods yet. Doesn't the Bible still prohibit charging interest for a money loan, regardless of what those clever medieval monks tried to claim to the contrary? You might think so. But let's look more carefully.

Two months after I graduated from college, I began an intensive course in biblical Greek at Asbury Seminary. Asbury is an evangelical seminary that grew out of the Wesleyan holiness tradition. The student body roughly reflected that tradition. Many of my fellow seminarians had attended Bible colleges that treated the Bible as the Word of God. They knew the Bible backward and forward. Unfortunately, some of them had memorized isolated texts without knowing the wider context of the passages within the particular book, or within the Bible as a whole. And very few had considered the original context in which those books were written. As a result, they often had memorized precise sentences but not precise meanings. Without intending to, they treated the Bible not as the unified Word of God, but as a collection of free-floating aphorisms, cut off from time, space, and history, like *Bartlett's Familiar Quotations*.

To help students overcome this bad habit, our professor, Bob Lyons, would frequently tell us, exaggerating for effect: "Context is everything." Take the word *no,* for example. If the lady at the ice-cream place asks me if I would like blue Superman ice cream, I'll say no. If a co-worker—*any* co-worker—asks me, "Do these jeans make me look fat?" I'll say no. But *no* in these two situations means different things. In the first case, it means: "No, I don't want Superman ice cream." In the second case, it means: "No, those jeans don't make you look fat"—or to the suspicious questioner, it probably means, "No, I'm not required to give you an honest answer." So the meaning of even a common word like *no* depends on the context it's used in. The same can be true of whole sentences. If a polite boy tells his teacher, "No, ma'am," when asked if he needs a drink of water, it means one thing. If a guy on the basketball court tells another guy, "No, ma'am," after swatting away his shot, it means something completely different.

The same thing is true with Scripture. The meaning of the words and sentences hinges on their historical and literary context—that is, when and for whom they were written, and in what genre. To take an obvious example, let's say you know nothing about the Bible, and I tell you: "The Bible says in Psalm 14: 'There is no God.'" I'm telling you the truth—sort of. The Bible really does have a verse with the words "There is no God." But what you would think I'm saying, and what I would be implying, is that the Bible teaches that there is no God, even though the Bible does no such thing. You'd realize this the second you looked at the context of those words in Psalm 14:

Fools say in their hearts, "There is no God."

All you'd need is four extra words to know you'd been snookered. The Bible doesn't claim that there is no God. It says that *fools* say there is no God.

This is an easy example. But the same lesson holds true for any biblical passage. You can't treat isolated biblical passages as if they were private little notes written just for you in your situation. Every passage is a part of a larger text, written at a specific time and place, and for a specific audience. When we read Paul's letters, for instance, we're reading mail from the first century. That doesn't mean we can explain away Paul's letters as historical artifacts. They're the inspired word of God. But it does mean that if we want to know their universal meaning—how they apply to us—we first have to figure out what they meant originally.

Sometimes this is pretty easy, once you think to do it. For instance, Paul and Peter both advise women not to wear braided hair or gold in church (1 Tim. 2:9–10; 1 Pet. 3:3–4). You might think that means that an American woman who wears braided hair in church in 2008 is disobeying Paul's advice. But think of the context. Paul is advising Timothy, a pastor of a church in first-century Ephesus. In that setting, apparently, wearing gold and braided hair were considered to be immodest. Wealthy women could spend hours having their hair elaborately braided and interlaced with gold jewelry. It was a way for them to show off their wealth.

In twenty-first-century America, braided hair is a simple, modest way to put your hair up, and gold jewelry is as common as bad grammar. So wearing gold and braiding your hair meant one thing in first-century Ephesus and means quite another in modern-day Yazoo, Mississippi. Paul wasn't pronouncing an eternal law against Christian women braiding their hair.

Now consider the biblical texts that prohibit charging for money. The Old Testament passages were all written when the Israelites had a static, agrarian, seminomadic society. In other words, they were written when not much new wealth was being created beyond what could be coaxed from the fields, and were written to people who didn't have much surplus money. The capital loan that rose to prominence in the Middle Ages is nowhere in sight. These passages are referring to rich Israelites lending money to their poor kinsmen for basic necessities. The Leviticus passage mentioned earlier makes this clear:

> If any of your kind fall into difficulty and become dependent on you, you shall support them; they shall live with you as though resident aliens. Do not take interest in advance or otherwise make a profit from them, but fear your

God; let them live with you. You shall not lend them your
money at interest taken in advance, or provide them food at
a profit. (Lev. 25:35–37)

Also, notice that this is describing how members of the faith
community were to treat *each other* (as well as resident aliens).
For some unstated reasons, Deuteronomy allows Israelites to
charge interest for money lent to non-Israelites. So neither text is
setting up a universal law against charging interest.

OK, but didn't Jesus up the ante? In a long sermon recorded in
Luke, Jesus said: "If you lend to those from whom you hope to
receive, what credit is that to you? Even sinners lend to sinners to
receive as much again. But love your enemies, do good, and lend,
expecting nothing in return" (Luke 6:34–35). Read in isolation,
this passage might lead you to think Jesus is prohibiting the charging of interest. But if you read the entire passage carefully (in light
of its historical setting), you see that something else is going on.

In the first part of this sermon, Jesus has given his famous
beatitudes, such as "Blessed are you who are poor" and "Woe to
you who are laughing now, for you will mourn and weep." Then,
in addressing "you that listen," he says, among other things: "If
anyone takes away your coat do not withhold even your shirt."
Does Jesus mean we should hope for everyone to be poor, so that
they can be blessed? Is he telling us not to laugh? Are merchants
disobeying Jesus if they sell shirts and coats? Is Jesus forbidding
society to enforce laws against theft? Not likely. The sermon is
deliberately provocative and hyperbolic—a common rhetorical
device in first-century Judaism. If you miss that, you'll miss the
point.

There's a lot going on in this sermon, but Jesus seems to be
reversing the popular belief that if you're poor, it's because God
has cursed you for disobeying him, and that if you're rich, it's because God has blessed you for obeying him. Here and elsewhere,
Jesus corrects that wrongheaded idea. He's also saying that if
we're his followers, we should go far beyond the usual standards
of generosity and forgiveness, even to the point of loving our enemies rather than hating them.

Now look again at what Jesus says about lending money. Jesus says that even sinners lend money, expecting to receive back the *same* amount. He says nothing about charging interest. Instead, he says we should lend *expecting nothing in return*. So he's admonishing gratuitous generosity, not denouncing banks for charging interest on business loans. Jesus's command not to expect repayment is no more a reproof of modern banking practices than his command to give someone the shirt off your back is a condemnation of clothing merchants who sell shirts.

Now look again at the other passages. When Jesus drives the money changers out of the Temple (John 2:13–22), he also drives out everyone selling sheep, cattle, and doves. He wasn't denouncing commerce or money changing in general, but protesting the misuse of a house of worship.[18] He may also have been protesting collusion between these merchants and the political authorities, who gave the merchants a monopoly.[19] If so, their commerce was unjust, no matter where it was located.

In his Sermon on the Mount, Jesus told his disciples: "Do not store up for yourselves treasures on earth, where moth and rust consume and where thieves break in and steal; but store up for yourselves treasure in heaven . . . for where your treasure is, there your heart will be also" (Matt 6:19–21). Here he's reminding his disciples that their ultimate loyalty is not to wealth or possessions, but to God's kingdom. He's not denouncing savings accounts and IRAs, but hoarding. At the time, burying was considered the safest way to hide money. But buried money is unproductive money. Money in a bank that earns interest (and is available for other ventures) is productive.

In fact, Jesus reserved some of his harshest words for those who seek false security in hoarding. Consider the parable of the talents (Matt. 25:14–30).

A man calls three servants and entrusts them with huge sums of money:

> To one he gave five talents, to another two, to another one, to each according to his ability. Then he went away. The one who had received the five talents went off at once and

traded with them, and made five more talents. In the same way, the one who had the two talents made two more talents. But the one who had received the one talent went off and dug a hole in the ground and hid his master's money.

You know the rest of the story. When the master comes back, he compliments and rewards the first two, but he lambastes the last: "You wicked and lazy slave! You knew, did you, that I reap where I did not sow, and gather where I did not scatter? Then you ought to have invested my money with the bankers, and on my return I would have received what was my own with interest."

This, like many of Jesus's parables, is about the kingdom of God.[20] Still, the parable contains a lot of economic wisdom. Notice that the first two servants are rewarded for investing the money they're given—for putting it at risk, where it can bear fruit. They're not praised for making money, but for being "trustworthy."

You might think the third servant was trustworthy; but the master punished him for playing it safe and hoarding. He expected the servant to invest, to put the money at risk. *At the least,* the master tells him, he should have put it in a bank where it could bear interest. If you're looking for Jesus's views on interest, this is the best clue there is. Jesus isn't giving an economics lesson—the parable is about the kingdom of God—but he would never have told this parable if he thought it was always immoral to accept interest for lending money to someone. On the contrary, he treats risk, investment, *and* interest in a positive light, and trusts his listeners to do the same. He describes enterprise as productive, not exploitative, and money as fertile, not sterile. Too bad Christendom didn't notice that sooner.

There's a larger lesson here. We must distinguish what the Bible actually says from what we assume it ought to say. Unfortunately, as we've seen, when it comes to economics, pious assumptions too often replace careful reading—and careful thinking.

Seven

Doesn't Capitalism Lead to an Ugly Consumerist Culture?

In 2007, I visited Hong Kong, a unique city-state covering several mountainous islands at the mouth of the Pearl River delta in southern China. Hong Kong started as a fishing village, and was a British Crown Colony from 1842 to 1997. It shared the benefits of many former British colonies: a stable rule of law without an overly meddlesome government. It wasn't utopia, of course, but it's rightly said that the British governed Hong Kong through the method of "benign neglect." The result is one of the greatest capitalist success stories in history.

In 1997, Hong Kong was handed over to communist China. So far, China hasn't messed it up. It is now home to over 6 million people, with more people arriving every day. Downtown is covered with towering glass skyscrapers, but across Victoria Bay in Kowloon, it's all stuff all the time. Urban malls and street markets with thousands of booths are dismantled and reassembled every day. Kitschy neon signs jut out from the sides of high-rise buildings, crowding the space between the street and the sky. And everywhere there are merchants, hocking Chinese souvenirs and Chinese massages; Gucci and Polo; Rolex, Giordano, and Hello Kitty; cell phones and fake Rolexes; fruits and meats you've never seen before, and some you wish you'd never seen.

For champions and critics alike, this motley scene is the essence of modern capitalism. For critics, at least, it's not a pretty sight, whether it's in Hong Kong, Houston, or Harrisburg, Pennsylvania. Many Christians hear "capitalism" and think urban sprawl,

strip malls, grasping merchants, homogenous suburban developments, ugly billboard signs, and a bland McDonald's hamburger
served everywhere from St. Louis to Shanghai, and everywhere
the insistent drumbeat: "Consume, consume, consume."

For all too many of us, the message comes through loud
and clear. In the United States, we eat so much that even our
poor people have problems with obesity. Sixty percent of us
are buried in consumer debt, paying Visa high interest rates
because we couldn't resist the annual shoe sale at Nordstrom
or the allure of a forty-eight-inch plasma TV. We toss out perfectly good computers and cell phones after two years, just because we want an upgrade. Heck, we even have *disposable* cell
phones. If we try to ignore the rat race, there's always an ad on
the radio or the side of the bus to entice us. "History will see
advertising as one of the real evil things of our time," predicted
Malcolm Muggeridge. "It is stimulating people constantly to
want things, want this, want that." He had a point. Great entrepreneurs, recall, don't just find ways to fill a market. They
anticipate what people would want if they knew about it, and
then go about creating brand-new markets and new demand.
Every Christmas, millions of children plead for some new toy—
Wii or a PlayStation or an American Girl doll—that didn't even
exist a few years earlier.

In response, many have started to seek the more intimate cultural traditions of the past. Some seek simplicity, buying locally
and eating organic food and free-range chickens. Others seek a
"slow food" lifestyle that attempts to restore the dignity of the
meal. And a few even return to the countryside as new agrar-
ians, giving up the fast-paced life of the modern city for a life
lived closer to the land. The responses differ and are sometimes
hard to follow; but the enemy is the same: the modern industrial, global, consumer-driven culture—the "commodification of
everything," as Jim Wallis puts it—which turns "all values into
market values, gutting the world of genuine love, caring, compassion, connection, and commitment for what will sell, for example, on a television show."[1] This is a motley list of complaints.
Let's try to tease them apart and consider them in turn.

CONSUMERISM

Almost everyone complains about consumerism. During Christmas 2007, the producers of *Super Size Me* got in on the action with a documentary called *What Would Jesus Buy?* It followed a fictional Rev. Billy as he traveled to malls with a singing gospel choir, badgering bemused shoppers. But what exactly is consumerism? In his book *Crunchy Cons,* Rod Dreher says that consumerism "fetishizes individual choice, and sees its expansion as unambiguous progress. A culture guided by consumerist values is one that welcomes technology without question and prizes efficiency. . . . A consumerist society encourages its members both to find and express their personal identity through the consumption of products."[2]

Unlike most of the arguments we've discussed, this one is as likely to come from the right as from the left. Dreher is a conservative. He asks: "How can you be a traditional-values conservative in a society whose very economic structure is designed to separate you, your kids, and your community from those values, and each other?"[3] Earlier conservative stalwarts like Muggeridge, C. S. Lewis, J. R. R. Tolkien, G. K. Chesterton, and Hillary Belloc asked similar questions. All worried that capitalism inevitably corrodes the very cultural values rightly prized by conservatives—faith, family, community—and replaces them with an obsession with stuff. Are they right? I think their argument identifies the symptoms of a real disease, but, unfortunately, it misidentifies the disease.

GLUTTONY

There's no sin in the Bible called "consumerism." But consumerism is similar to one that is in there—gluttony. It generally refers to eating and drinking too much, but applies to the overconsumption of other goods and services as well. Proverb 23 warns:

Do not be among winebibbers / or among gluttonous eaters
of meat;

for the drunkard and the glutton will come to poverty, and
drowsiness will clothe them with rags.

The Christian tradition lists gluttony as one of the seven
deadly sins. Gluttony is not mere consumption. We have to
consume food, water, vitamins, clothing, and all sorts of other
things simply to survive. There's nothing wrong with this.
Eating, drinking, breathing, building, lighting fire: the Bible
treats these as the stuff of life from beginning to end. John describes Jesus eating fish for breakfast after his resurrection. Some
acts of consumption, such as eating bread and drinking wine in
Communion, are even means of grace.

Gluttony is a problem of excess, of not restraining our ap-
petites when we should. But how do you know excess when you
see it? Most people think of obesity when they hear gluttony. But
a person can be obese because of a slow metabolism or thyroid
problems, and a skinny person with a fast metabolism could be a
glutton and you couldn't tell.

So how much is too much? When I ask groups this question, the first answer is always the same: it's excessive to own
or consume things that go beyond our needs and merely fulfill
our wants or desires.[4] If that's right, we're all in serious trouble,
since, in modern societies, most of what we use we don't need
to survive. Just looking around the room where I'm writing,
I can find hardly anything I need to survive: a Christmas tree
with white lights, a big dining-room table with a runner and
place mats, a hutch, pictures and plates on the wall, a cool lamp,
the light fixture over my head, a little Eskimo Nativity set, and
the laptop I'm typing on. I do need shelter, since it's snowy and
twenty-two degrees outside. But I could probably get by without
the crown molding, curtains, and fancy electrostatic air filter. Is
all this a sign of excess?

Not necessarily. The problem is that there are no strict criteria
for determining where needs leave off and mere wants begin. Of
course, we can tell the difference at the extremes. My need for
air and water is different from my desire for the most powerful
video iPod currently on the market. Since as far back as Aristotle,

moral philosophers have recognized the difference between absolute needs and relative wants. If a person is generally healthy over the long term, and isn't suffering severe weight loss or frostbite or illness, he's probably getting what he needs: enough calories, nutrients, water, air, sleep, shelter, and so forth. But it's well known that infants in orphanages can receive all these things and still fail to thrive for lack of love and tenderness. So some amount of love is a need. Still, does anyone really think that if he gets more air, food, water, shelter, and love than he needs to survive, he's a glutton?

Besides, we have many needs that don't bear on our survival. To do my job, for instance, I have to ride on an airplane, type on and use a computer, have e-mail and cell-phone coverage, dress in a certain way, and so on. Take this stuff away and I won't suddenly drop dead. But I won't be able to do my job. So needs vary with time, place, and circumstance. Of course, I could get by with a cheaper computer or cell phone or suit, but there is no black line that separates need from excessive desire. It's not like a $299 suit is fine but a $300 suit will get me a ticket to hell. The issue is more subtle than that.

Moreover, there's no biblical rule that requires us to consume only what we need to survive. The Bible has a lot to say about the dangers of wealth and overindulgence, but it also frequently treats bounty as a sign of God's blessing. In the right context, prosperity and abundance even foreshadow the blessings we will enjoy in the kingdom of God, a land "flowing with milk and honey," where, presumably, we will never fear for our survival.[5]

The Bible doesn't condemn feasting. In fact, the Old Testament actually mandated certain feast days, and Jesus apparently enjoyed the occasional party enough that he was falsely accused of being "a glutton and a drunkard" (Luke 7:34).

So neither abundance nor enjoying more than is strictly necessary is the problem.

Gluttony involves the state of our hearts, the orientation of our desires, our priorities. Saint Augustine talks about the "beautiful form of material things," which, while beautiful, can be an occasion for sin: "Sin gains entrance through these and similar good things when we turn to them with immoderate desire, since they are the lowest kind of goods and we thereby turn away from the better and higher: from you yourself, O Lord our God, and your truth and your law. These lowest goods hold delights for us indeed, but no such delights as does my God, who made all things; for in him the just man finds delight, and for upright souls he himself is joy."[6]

"Immoderate desire" is the point, not consumption. Jesus said as much when he warned that we cannot serve both God and Mammon (Matt. 6:24). Paul made much the same point when he said that the *love* of money (not money itself) is the root of all kinds of evil. You can have only one object of ultimate loyalty and worship. God is the only proper object of worship. So however good something is in itself, if it crowds out God, it's an idol.

Notice that gluttony isn't the same as envy, since you could still be consumed by your desire for stuff, even if you were perfectly happy with what you had and perfectly happy for other people to have what they want as well. But that means that since gluttony is a state of the heart, a poor person could have misplaced loyalty to, say, cheap knickknacks while a wealthy person could be wearing expensive jewelry and still have her priorities straight.

That said, gluttony can have outward expressions. In 1899, a guy named Thorstein Veblen wrote *The Theory of the Leisure Class*. He argued that in modern societies, the rich often spent money simply for the purpose of showing their social status. He called this strange phenomenon "conspicuous consumption." Although he overworked the concept, Veblen was surely onto something. We've all seen pictures of Hollywood divas wearing designer dresses of no particular distinction except that they cost as much as the average taxi driver makes in a year. The significance of the dress is in its high price, not its beauty, because it elevates the woman's social status. You don't have to have overly refined moral sensibilities to find that sort of thing distasteful.

Conspicuous consumption is consumerism on steroids. But consumerism manifests itself in a million smaller ways. It's a serious moral problem when it involves a misplaced loyalty to material things—to fine food, wine, clothes, cars, iPods. It often involves our seeking ultimate happiness in our possessions. Some studies suggest that once people have their basic necessities met, there's not a tight connection between one's possessions and one's happiness. The fact that you make twice as much as your second cousin doesn't mean you're twice as happy. Stuff can never be our ultimate source of happiness.

Unfortunately, judging from the number of people who find themselves neck deep in consumer debt, too many of us are still trying to find fulfillment in stuff. In October 2007, there was some $17 billion at least one month overdue to America's seventeen largest credit-card companies. That was up from about $14 billion the year before.[7] That's a lot of people living beyond their means. If you have this problem, get help. A good place to start is with the books or Web sites of Dave Ramsey and Ron Blue. But I'm not offering financial advice in this book. Here, I just want to know whether capitalism and consumerism have to be two peas in a pod.[8]

We've already seen in chapter 5 that despite the stereotype, Americans are more generous with their charitable giving than people in other industrialized countries. But rather than trying to figure out whether Americans are more gluttonous than their counterparts in preindustrial or socialist societies, let me just offer a few simple distinctions that I think will do far more to shed light on this subject.

Capitalism refers to an economic system with rule of law and private property, in which people can freely exchange goods and services. Free exchanges, by their very nature, will be viewed as winning exchanges by all parties involved. Otherwise the free parties wouldn't be involved in the exchange. So capitalism enhances freedom and channels even our baser instincts (like greed) into the task of creating win-win scenarios with the people around us. A free market is best for distributing goods, services, and information, whether they are trucks, trumpets, or trashy novels. But the system doesn't determine what choices people will make.

Too many critics confuse the free market with the bad choices free people make. Rod Dreher, for instance, chastises fellow conservatives, saying, "We look down on the liberal libertine who asserts the moral primacy of sexual free choice, but somehow miss that the free market we so uncritically accepts exalts personal fulfillment through individual choice as the summit of human existence."[9] Perhaps they miss that fact because it's not a fact. The free market doesn't exalt anything. Human beings exalt and denounce things like sexual free choice. Human beings might exalt "individual choice as the summit of human existence," but a system of free exchange doesn't do that.

In a free economy, sinful entrepreneurs may entice customers with pornography, and sinful customers may buy it. But having free choices in the market doesn't dictate what people will choose. That's the whole point of freedom: it always involves costs—that is, trade-offs. To choose one path is to foreclose the opposite path. Even God accepted trade-offs. He chose to create a world with free beings, one that allowed those beings to turn against him. And they did. But their freedom didn't cause them to choose the bad. It just allowed them to. So, too, with a free economy. Critics notice all the vice present in free societies. But it is only in free societies that we can fully exercise our virtue. Charity is charity, for instance, only if it's not coerced.

Besides, there's no evidence that state control of the economy makes a citizenry more virtuous. Every social ill in modern-day America, from widespread abortion and alcoholism to family breakdown, was much worse in statist and communist countries.

That suggests that something other than mere economics is at work. We shouldn't expect the economy, free or otherwise, to instill virtue in people. While a free economy requires and may even encourage certain virtues, like trust and basic honesty, and while it may channel sins like greed into productive behavior, by itself it can't teach us God's complete will for our lives. It can't raise up sons and daughters in the way they should go. It can't take the place of the family, church, synagogue, and Boy Scouts.

A free economy requires not only political freedom, but also free institutions of civil society, to produce a virtuous people.

Contrary to the stereotype, capitalism is not compatible with a vicious populace. Consumerism in particular is actually hostile to capitalism, at least in the long term. Thinkers as diverse as Max Weber and Karl Marx understood that capitalism and overconsumption could not long coexist. In his famous book *The Cultural Contradictions of Capitalism*, Daniel Bell worried that the "Puritan" culture of restraint that gave birth to capitalism had been lost, leaving a culture of hedonism where mere pleasure has become the purpose of life.[10] Left unchecked, this hedonistic culture, he predicted, would destroy capitalism.

Remember, advanced capitalism needs financial habits and institutions that allow wealth to be reinvested so that wealth itself becomes wealth producing. That requires not only that wealth be created, but that some of that wealth be saved rather than consumed. Delaying gratification is restraint; it's the opposite of gluttony. So consumerism is *hostile* to capitalist habits and institutions. This is why statistics about consumer spending are not reliable indicators of the long-term health of an economy. Every economy will have consumption. But a sustainable capitalist economy needs large portions of wealth creating, saving, and investing as well.

YEAH, BUT IT'S STILL BAD

But once we've distinguished the sin of consumerism from capitalism itself, there remains an assortment of vices attributed to capitalism. The accusations have poured from the pens of intellectuals for almost two centuries now—that capitalism alienates us from nature, that it destroys a "traditional way of life" and local cultures, that it leads to ugly and crude art, music, and architecture.[11] "The free market extolled by conservatives as the holy of holies is destroying communities, and turning us all into slaves of the economy," says Rod Dreher in *Crunchy Cons*.[12] Such complaints are so various, vague, and sometimes contradictory that responding to all of them would take a book unto itself; but let's consider a few of the really popular ones.

GOING LOCAL

I happened to work in downtown Seattle during the 1999 WTO protests/riots. I took the opportunity to ask a random sample of participants (most of whom claimed to be opposed to "global-ization") what they were protesting. I got a variety of answers, but they all shared a common theme: that global capitalism is evil. It destroys indigenous cultures, family farms, mom-and-pop stores owned by and employing your neighbors, along with all the interesting flavor of diverse places, cuisines, habits, and cultures. It replaces them with big, monochromatic Walmarts, Home Depots, and strip malls. And it feeds all of those evil mul-tinational corporations. According to some, the multinationals homogenize the global economy, sucking the variety out of it. According to others, they provide too much variety, a spiritually debilitating cornucopia of consumer choice. Some protesters held both views at the same time. I came away from the WTO pro-tests with the sense that most of the protesters were really there for a big parade and had almost no idea what they were talking about.

If you want to hear a well-articulated case against global capi-talism, you'll need to look somewhere else, such as the seminal group of essays by economist E. F. Schumacher called *Small Is Beautiful*.[13] Schumacher's argument, in a nutshell, is that modern capitalist economies are unsustainable, since they produce larger and larger industries that use up more and more nonrenewable resources. Such economies also create dehumanizing workplaces for most people, he argued, workplaces that should be replaced with smaller, localized, "human-sized" work environments, like villages, that are more dignified and sustainable.

Schumacher's argument has continued unabated to the present, trumpeted in books such as *Deep Economy*, by Bill McGibben. McGibben argues that the "Wal-mart Nation" is unsustainable and needs to be replaced by a human-scaled, sustainable "Farmers' Market Nation."[14]

Given this constant drumbeat, many are now convinced that they have a moral and economic obligation to buy locally and avoid stores owned by big companies outside their community.[15] Suddenly, provincialism is progressive.

Sometimes it makes sense to buy locally. In the summer, I can get the best vine-ripened tomatoes at the Fulton Street farmers' market about a mile from my house. But applied to everything, the advice to "buy local" ignores most of the lessons of basic economics. If correct, the United States would be better off if we had stayed separate colonies trading goods, services, and capital only within the borders of each colony. And we'd be better off if the individual cities just traded within their borders, and still better off if the neighborhoods—wait—individual families—wait, individuals—just traded with themselves.[16]

Only a few lone survivalist nutcases try to take the local-is-better philosophy to such an extreme, but that's the logical end of the illogical train. Stopping the illogical train at the state or county level only masks the irrationality, and that only very thinly.

> "What is prudence in the conduct of every family can scarce be folly in that of a great kingdom. If a foreign country can supply us with a commodity cheaper than we ourselves can make it, better buy it of them with some part of the produce of our own industry, employed in a way in which we have some advantage."
> —Adam Smith

If you remember the trading game, you know that the larger the arena of free trade, the better off are its participants. Six billion people trading only with themselves, or only with their families or closest neighbors, would not create 6 billion self-sufficient people. It would create several billion dead people, and leave the rest to eke out a meager existence as subsistence farmers, hunter-gatherers, or scavengers. Division of labor, in which individuals, cultures, and countries can specialize in areas that give them a competitive advantage (that is, in what they can do best, or at least better than many others) and trade for other things, has created vast wealth and allowed billions of people to live who would otherwise have died or never been born.

Although the now-developed world has benefited the most, even the poorest parts of the world have gained dramatically from industrialization and globalization. As economist Martin Wolf notes, "Africa's average real income per head is perhaps three times higher than it was a century or so ago. Asia's as a whole is up six-fold since 1820 and Latin America's nine-fold. In 1900, life expectancy was a mere twenty-six in today's developing world. It was sixty-four in 1999. This is much the same as the sixty-six achieved by today's advanced countries as recently as 1950."[17]

One could fill a thousand pages with counterexamples to the "buy locally" mantra. Imagine if the Danish closed their borders. Where would they get citrus fruit? If those of us who live in Michigan bought only locally, we wouldn't have fresh produce for most of the year. Regions with droughts or floods would starve to death. People far from salt mines would get goiter. Japan and Hong Kong would have very little of anything. Third-world coffee farmers and factory workers would lose all their wealthy customers in the developed world. The locals would suddenly have more coffee, bananas, furniture, rubber, and tea than they knew what to with, at least until most of the producers of those products went out of business for lack of paying customers. Eventually, the prices of most things would skyrocket, for rich and poor alike.

Few of us have any idea how much we, and others, benefit from global markets. Take a trivial example: a man's suit. Recently a Philadelphia designer and Drexel University professor named Kelly Cobb decided to see what it would take to design a suit (with tie, socks, and shoes), all with materials made within one hundred miles of her home. She had the benefit of a highly populated area that combined cities, farms, and ranches. It was still a Herculean effort. "The suit took a team of 20 artisans several months to produce—500 man-hours of work in total—and the finished product wears its rustic origins on its sleeve."[18] It looked like something the town drunk from all the old Westerns might have worn to his day in court.[19]

And Cobb estimated that, despite the artisans' best efforts, about 8 percent of the outfit still came from places outside the hundred-mile radius. In contrast, you can get beautiful tailor-made suits from Hong Kong in three days, from measurement to finish. The tailor focuses on his competitive advantage—measuring and tailoring—and imports the fabric from Australia. Although such personalized suits aren't cheap, they cost a tiny fraction of the locally grown suit from Philadelphia and look a lot better. For that matter, decent, mass-produced suits for one hundred to three hundred dollars look a lot better, and these can be found in any American city. And we're talking about suits here. Think if we were running the same experiment with cars or cell phones or medical imaging equipment. . . . You get the idea.

BIGGER IS BETTER—EXCEPT WHEN IT'S NOT

Affordable suits (or cars or cell phones or imaging equipment) are possible not only because of division of labor, but because of "economies of scale." If a company can find one hundred thousand customers, it's worth the company's investing in sophisticated "labor-saving" equipment so that it can mass-produce the product. If a company produces just six cars, the costs per car will probably be astronomical, as they were in 1896. But then Henry Ford pioneered mass production in the early twentieth century by using assembly lines. Ford could sell cars at prices that ordinary, middle-class Americans could afford because he so reduced the cost of production for individual cars.[20] Automation and mass production make exquisitely complex technologies like DVD players and cell phones available to almost everyone in modern societies, technologies that initially were available only to the very rich.

Large chain stores like Walmart often replace local mom-and-pop stores. It's not because they're part of some evil globalization conspiracy, but because they enjoy greater economies of scale. They can buy, sell, and distribute in bulk, and can negotiate lower prices with suppliers because of their purchasing power. If you produce music CDs, and Walmart offers to buy a million of them, you can afford to sell the CDs near your cost of production and still make a handsome profit. The size of a Walmart or Target allows it to sell many products cheaper than the local store does.

But no one is forced at gunpoint to shop or work at Walmart. Given the choice, many people prefer the savings of a Walmart or a Target to whatever virtuous feelings that might accrue to them by paying more and getting less at local mom-and-pop stores, which often enjoyed near-monopolies in small communities before the competition moved in. It's easy to forget that even Walmart started out as a local store in Rogers, Arkansas, owned by Sam Walton. It slowly grew—not through a pact with the devil, but because customers preferred it to the competition. In fact, the company didn't get really big until Walton was in his fifties. It now employs over a million people worldwide, making it the largest private employer on the planet. None of those employees are forced to work at Walmart.[21]

So does that mean that in a global economy small operations will always give way to giant, multinational corporations? No. Just as there are economies of scale, so too are there what are called *diseconomies of scale*. Larger companies are generally more bureaucratic, top-heavy, and slower to react to radical changes. "What small companies give up in terms of financial clout, technological resources, and staying power, they gain in flexibility, lack of bureaucracy, and speed of decision-making," observed one Indian entrepreneur.[22] That's why new companies spring up all the time (Google started only in 1998) and big companies collapse (think Enron). GM is the largest car company in the world, but it's not the most efficient or profitable.

Bigger is better for some things, but not everything. Small, local charities, for instance, tend to help vulnerable populations with long-term problems more effectively than big, impersonal organizations. And ideal size varies from industry to industry.

The best size for an elementary school is not the same as the best size for a coffee shop or an oil refinery or a gas station.

Moreover, even though Walmart, Costco, and Starbucks are giant corporations, their individual stores are only so big. They don't just keep getting bigger and bigger. And Walmart doesn't ship unknown employees from Arkansas, or robots from Taiwan, to staff its stores. The same local people, with local knowledge, who would work and shop in a mom-and-pop store end up in the Walmart as well. There are two coffee shops near my house—Starbucks and Kava House. I'm just as likely to see people I know at the Starbucks as at funky, locally owned Kava House. And the Starbucks is actually the smaller and cozier of the two.

Sometimes local knowledge outweighs economies of scale, giving small stores an advantage. And sometimes local insights are then expanded or bought out (at market price) by larger operations. Social worker Stacy Madison started with one business partner, Mark Andrus, with a single sandwich cart, selling pita sandwiches in downtown Boston. The cart got so popular that they started baking the pitas into little pita chips so customers would have something to munch on while they waited. Sales of these healthy alternatives to Fritos grew along with the health consciousness of yuppies. And before long, a company named Costco started selling the chips. Now you can get them at Target, too. Stacy's is the number-one pita-chip brand in the country.[23] Nobody had heard of them a few years ago. American business is filled with such stories. In market economies, most little operations get big not by getting evil, but by serving customers.

FACTORY FARMING

Yes, but what about the alienating effects of big business? The poster child for this objection is "the factory farm," a vaguely understood institution that conjures up images of giant, government-subsidized, corporate-owned farms with animals packed together like sardines, jacked up on steroids and antibiotics, moved to and

fro by robots and conveyor belts, and shipped off to the slaugh-
terhouse without ceremony. "Factory farming" is almost always
blamed on capitalism.

Most of the confusion here results from muddling separate
issues. First, cruelty to animals can be a problem anywhere,
whether a society is capitalist, communist, feudalist, or prefeu-
dal agrarian. Second, government subsidies aren't capitalism-
in-action. They're just the opposite. Right now, corn subsidies
are popular, because corn is used to make ethanol, which is
mistakenly thought to be a viable alternative to oil.[24] Subsidies
often give large farms an advantage. Subsidies distort market realities, encourage overproduction, and make it hard for farmers
in developing countries to compete with American agriculture.[25]
There are any number of creative and, I think, bogus arguments
in favor of farm subsidies, but the point to remember here is that
government subsidies for growing (or not growing) a crop are
not the outcome of the free market. Too many critics of capitalism miss this fairly elementary point.[26]

Third, the fact that a farm is large and uses modern technology doesn't make it evil.[27] In December 2007 I visited a large
turkey farm outside Zeeland, Michigan, owned by Glenn Overweg.
Overweg, who lives with his family on the farm, is one of sixteen
members of the Michigan Turkey Producers Co-op. The co-op
raises, processes, and cooks some 4.5 million turkeys a year. At
any one time, the Overweg farm is raising sixty thousand turkeys in three enormous barns. Several times a year, they receive
twenty thousand tom chicks from Canada and house them in
the "brooder" barn for four weeks. It takes three people an hour
to unload the new batch of chicks. Chicks need about a square
foot each to prosper, which determines the size of brooder barns.
That might not sound like much room, but it's plenty for a tiny
chick.

The barn is like a giant incubator. On the day I visited, there
was a snowstorm. It was twenty-five degrees outside. Inside, it was
a cozy ninety degrees. The yellow chicks eat, drink, sleep, and
grow in the giant brooder barn for four weeks and are then

moved to the "grower" barn. After they fatten up a bit more, they are moved to the "finisher" barn, where each turkey has three and a half square feet of space. There, most of the turkeys will grow to thirty-eight to forty pounds (at around nineteen weeks of age). Then they will be processed and cooked in a facility in Grand Rapids and shipped all over the country and as far away as Mexico. If you ever eat Boar's Head turkey meat, you've probably benefited from the Overweg farm.

Glenn Overweg recently built another specialized two-hundred-thousand-dollar barn. In the last several years, selective breeding has produced domestic turkeys that grow so fast they don't need three barns. So Overweg will soon go to two barns to keep his operation competitive. (Only one of his original barns will work with this new method.) When I asked him about cost pressures, he said the next two years would be tight because of the high price of corn, created by—you guessed it—corn subsidies and the ethanol craze.

No doubt some would call this a factory farm. The turkeys certainly aren't freely roaming in a field. (If they were, they'd be frozen solid.) It's not as pretty as, say, a flock of white sheep on a chartreuse New Zealand hillside. But what is? Still, there's nothing inhumane or alienating about the Overweg turkey farm. It's a monument to human ingenuity and hard work.

As mentioned in chapter 4, methods like those Glenn Overweg uses mean that only 1.9 percent of the American population now lives on farms. And yet that small percentage produces enough not only to feed the American population, but to export abroad. Far fewer people *have* to live the "traditional agrarian lifestyle" (to quote Rod Dreher). If that sounds like a bad thing, it's probably because most of us have no idea what that lifestyle involved. Let's put it in plain language: billions of people aren't stuck working sixteen-hour days on a farm when they'd much rather be doing something else—something they just might do better. In developed countries, most people are able to spend more of their time doing things not related to feeding themselves—like painting, listening to music, discovering extrasolar planets, going to movies, symphonies, and NASCAR races, sledding with children, or even buying a home on two or three acres and enjoying some small-scale farming on the side.

People in advanced economies sullied by "factory farms" also spend less and less on food as a percentage of their income, even as they enjoy a more diverse diet. Surely, on balance, this is a good thing.

TABLE 2. EXPENDITURES FOR FOOD, 2004: COMPARISONS BY COUNTRY

Country	% of household expenditures used for food	Expenditure per capita on food (US$)
United States	5.7	1,592
United Kingdom	9.1	2,054
Canada	9.7	1,565
Australia	10.6	1,820
France	14.9	2,709
Colombia	26.2	365
Thailand	29.2	405
Philippines	37.7	268
Jordan	46.2	604
Nigeria	54.3	252
Azerbaijan	57.3	247

Source: U.S. Department of Agriculture, Economic Research Service, http://www.ers.usda.gov/Briefing/CPIFoodAndExpenditures/Data/2004table97.htm

LIFE IS BEAUTIFUL

OK. But economics isn't everything. I've focused on the economic benefits. What about the costs? Rod Dreher puts it starkly in his "Crunchy Con Manifesto": "Beauty is more important than efficiency."[28] When the little downtown stores close up and big malls open on the edge of town, doesn't the quality of life go down? When agribusiness replaces a patchwork of family farms, isn't something lost? What of the community that existed before?[29]

Yes, in these cases, something—perhaps something valuable—is lost. Only 1.9 percent of Americans now live on farms. So one cost is whatever aesthetic and spiritual benefit farm life (as celebrated by great writers such as Wendell Berry) could provide to the other 98.1 percent. But this is part of a trend that's lasted thousands of years. At some point in the past, hunting and gathering gave way to farming in much of the world. Even farming has had several incarnations. For a long time, most civilizations were based on slave labor. The slaves did most of the farming. In Europe, slavery eventually gave way to feudalism. The serfs did most of the farming. In the West, feudalism eventually gave way to capitalism (though slavery reemerged in the colonies). By modern standards, the farms of the American colonial era worked by freemen were small. Over time, technology and economies of scale led to much larger, more efficient farms. Just in the last fifty years, the world has undergone the so-called Green Revolution. With a combination of fertilizer, high-yield grains, and industrial farm techniques, we've been able to produce far more with far less labor, and to feed billions more people than before.

Let's return to the claim that beauty is more important than efficiency. This begs the question, More important for what? Beauty may be more important than efficiency if you're preparing a ceremonial wedding feast. But efficiency is much more important if you're trying to deliver food and water to an Indonesian town decimated by a tsunami. In that case, nobody cares about the flatware or presentation.

But let's say beauty *is* always more important than efficiency. It's still a bad argument against capitalism. Modern industrial farming techniques have made it possible for billions of people to live who would otherwise have died, especially in third-world countries like India. A family having food on the table is a beautiful thing. Human beings are the most beautiful things in all of the physical creation. Remember, humans alone are made in the image of God. So if beauty is more important than efficiency, we should welcome more efficient methods that allow more wonderful human creatures to live. Any aesthetic judgment that ignores the beauty and worth of human life is blinkered.[30]

THE ONLY CONSTANT

Change, it's often said, is the only constant. What is disconcerting about contemporary life is the rate of change. Entire communities can be built around an industry—gold mining, coal mining, horse buggies—that can become obsolete overnight. Economist Joseph Schumpeter described this effect of modern capitalism as "creative destruction."[31] When entrepreneurs figured out how to retrieve and extract energy from petroleum, the whale population may have breathed a collective sigh of relief, but the whaling life—the skill, the camaraderie, the derring-do—was largely destroyed.

In most places, oral traditions eventually gave way to writing. Later, ordinary writing was transformed by the printing press. As great a blessing as writing and the development of affordable printing have been to human culture, even these changes came with trade-offs. People in oral cultures tend to memorize much more than people with written languages; but only a very foolish brand of nostalgia would wish all the printing presses destroyed and the secret of writing forgotten. The real question, here as with all technological innovations, isn't whether there is a downside. There almost always is. The question is whether the overall benefits outweigh the costs.

It's easy to wax nostalgic about the "sacramental" value of working with the land to produce food (to quote Rod Dreher again) and to conjure up an image of Hobbits peacefully tending their crops in the Shire. Appreciating the genuine good in such work is itself a good. But nostalgia is another matter. The danger with such nostalgia is seeing in the past only benefits, and in the present only costs.

Any fair comparison must count costs and benefits, and face the live alternatives. Local communities built up around agriculture or auto manufacturing are not eternal realities that must be preserved at all costs. Think of coal mining in West Virginia, which right now provides some forty thousand jobs. Even though there are 117 fruitful coal seams in the state, at some point, large-scale coal mining in West Virginia will become obsolete. Long before we run out of coal, it will become so inaccessible and expensive that it won't make sense to use coal for energy. And this moment will, in all likelihood, be hastened by new sources of energy and energy retrieval. (For more on this, see chapter 8.) When that happens, towns formed around coal mines will have to change or will disappear when people move elsewhere.

Writers will bemoan the loss of the traditional mining way of life. But what's the alternative? Would it be better for people (including coal miners) to be *forced* to pay for coal-based energy, even after it gets really expensive or has been replaced by new technology, just to preserve these communities? Would it be dignified for coal miners to keep scraping isolated bits of coal out of the ground, knowing their work was propped up by a government handout, work that produces something no longer needed? Of course not. As challenging as the transition will be, these workers should reallocate their labor to more productive uses for which there is demand.

THE VILLAGE RETURNS

In capitalist societies, some communities disappear even as new ones appear. There's no reason to deny that. But it sounds worse than it is, since many types of communities reappear in the more prosperous societies. After all, in many cities, small, locally owned stores survive or reemerge when enough people prefer them, even if they're more expensive. Seattle neighborhoods like Wallingford and Fremont and the University District have little town centers with farmers' markets and Pagliacci pizzerias (local) clustered near places like Starbucks (well, in Seattle, Starbucks is locally owned, but you get the idea). Even downtown shopping often reemerges in larger cities. In New York, Seattle, and Chicago, big malls coexist with hip downtown shopping areas, each serving its own, sometimes overlapping, constituencies.

Similar patterns exist even in smaller urban settings. During the summer, we partner with another family here in Grand Rapids, Michigan, to buy our produce from a "communal sustainable agriculture" operation—all local, all organic. All summer, we pick up the produce at a designated location, packing it into canvas bags to avoid the paper-or-plastic dilemma. Every week brings another mystery green, and by the end of the summer we've sampled up to seven varieties of eggplant. (I'm still looking for one I like!)

How is this possible in an era of farm subsidies and big agribusiness? Answer: demand. That's the beauty of the *free* market. If enough people have enough disposable income to buy, say, locally grown organic green goddess eggplant rather than the outsized purple eggplant at the nearby supermarket, entrepreneurs find a way to provide it. You want eggs from free-range, organic chickens raised on flaxseed that cost three times as much as regular eggs? If enough people share your proclivities, it won't be long before you find a place to buy them.

Ever heard of Trader Joe's? Whole Foods? These store chains didn't spread because of U.S. Department of Agriculture programs to preserve the "natural-foods lifestyle." They spread because of consumer demand in a free market. Organic products have gotten so popular that even Target stocks organic eggs and soy milk. The Michigan Turkey Producers Co-op now includes some organic turkeys in its portfolio. This is just what you would expect in a market economy.

But what if such a lifestyle isn't organic enough for your tastes? Well, if you have enough disposable income, you can buy a small farm, raise your own livestock in the way that suits your culinary proclivities, and commit yourself to the "eco-gastronomy" of a "slow food" lifestyle.[32] Modern, industrial capitalism has allowed even the poor to eat and live well by historical standards, even as it gives many of us the freedom and prosperity to indulge in such luxuries.

STILL UGLY

OK, OK. But what about all the ugly suburban McMansions sprawling across the once-pleasant countryside? What about the tools of industry that scar the landscape? What about our hurried "fast-food nation"? What about the misogynist bilge of 50 Cent and degrading reality TV like *Wife Swapping?* What about gas-guzzling, beer-chugging, bleached-blondes-in-tight-T-shirts-hollerin', seven-car-pileup-lovin' NASCAR? Huh? Am I really going to defend all that?

Well, if we're talking about *my* preferences, then no. I prefer fresh, whole produce rather than frozen or processed food. I'd much rather savor a high-quality, healthful, home-cooked meal with friends and family than wolf down a Quarter Pounder with cheese in my car. I prefer Beethoven and Bach to Britney Spears (though I do like the Barenaked Ladies). I don't like to look at the ugly factories in Gary, Indiana. I prefer old houses, in older urban neighborhoods with sidewalks and large trees, to new suburban developments with big new houses and little trees. And at the risk of getting tarred and feathered by my childhood friends back in Amarillo, I even prefer *The Nutcracker* to NASCAR. Moreover, I'm not a relativist. I think that a Mozart symphony is an objectively better form of entertainment than, say, mud wrestling.

Other than my indifference to Birkenstocks, I share many preferences with "crunchy cons" and other critics of modern capitalism on the left and the right.[33] But, unlike most of those critics, I see these preferences for what they are: luxuries.

To his credit, Rod Dreher tries to anticipate this response. In the more nuanced version of his "Crunchy Con Manifesto," he replaces "beauty is more important than efficiency" (which is on the book's back cover) with this: "Appreciation of aesthetic quality—that is, beauty—is not a luxury, but key to the good life."[34] Sounds nice, but what does it mean? It seems to me that something could be both a luxury and key to the good life. In any case, obviously people differ in their aesthetic discernment. Some people are color-blind or tone-deaf. Others have so many worries about basic necessities that they can't cultivate the refined eye required to distinguish, say, Claude Monet from Thomas Kinkade.

Besides, highbrow choices are often expensive choices. Appreciating them might not be a luxury, but satisfying them often is. A unique, pristine, eighty-year-old brick Tudor Gothic house with mature landscaping in a historic neighborhood protected by finicky zoning laws costs a lot more than one of twenty cookie-cutter houses with plastic siding built in a sprawling new development on the edge of town.[35] Even if you can tell the difference, it doesn't mean you'll have the money to buy the Tudor Gothic. Or, you might have the money and the discernment but have other priorities. Perhaps you would rather save for braces for your five beautiful kids than indulge your taste for architectural detail.

Most of us make the same kind of calculations with, say, fast food. Sure, it's not the healthiest or most meaningful experience you'll ever have. You shouldn't eat it every day. But sometimes you need a cheap meal fast so you can get on to other things, like being there for your kid's soccer game. These are the kinds of ordinary choices we make every day. Sophisticated critics posture and overinterpret them. Sometimes a cheeseburger is just a cheeseburger.

In *The Anti-Capitalist Mentality,* the great Austrian economist Ludwig von Mises points out that the literati often make the mistake of comparing the furniture, art, and architecture available to the aristocracy in previous centuries with mass-produced things that are widely available to the average person now.[36] Capitalism gives many people—the great unwashed—the means to make choices denied earlier generations. By definition, the great unwashed don't always make refined choices. Lots of people think Domino's pizza makes for a heck of a good meal. They have neither the time, nor the budget, nor the inclination to find and frequent the local gourmet Italian restaurant. But isn't it better that they can enjoy Domino's pizza in heated homes than have to live a subsistence existence where such luxuries are a fantasy? That's the relevant comparison between past and present.

Critics complain about vulgar popular culture. Much of it is vulgar, of course, but the vulgarity isn't unique to capitalist countries. And it's easy to miss the hundreds of well-funded literary societies and libraries and art galleries, the opera houses in larger cities, the theater companies in midwestern towns, and the symphonies in cities everywhere. In previous centuries, only aristocrats enjoyed such pleasures. Capitalist wealth has made them available to vastly more people.

Still more significant, in a capitalist economy, refined but costly choices often become widely available later.[37] For instance, *billions* of people can now learn about and enjoy Mozart, Handel, Shakespeare, Puccini, and Bach because of mass-produced printing, recording, and communications technology.[38] And then there are new art forms, from film to computer graphic design, that rely on recent technology. Almost everyone has access to them. Louis XVI could never have imagined such luxury.

Sure, some things are just ugly. I don't know anyone who thinks ordinary strip malls and oil refineries are beautiful. But that's not their purpose. It's unreasonable to expect every building to satisfy high standards of artistic merit. Uniform aesthetic laws for, say, grocery stores would jack up the price of many goods and services, making them out of reach for those on the bottom rung of the economic ladder. Sometimes surface beauty takes a backseat to other concerns. That's OK.

The real problem is not strip malls and factories that serve legitimate purposes, but modern architecture, art, and music that *celebrate* meaninglessness and ugliness rather than truth, goodness, and beauty. Capitalism is *not* the culprit here. The culprit is the materialist worldview that has infiltrated almost every nook and cranny of Western culture, a worldview that insists that "beauty" has no objective basis in reality. Wherever we find the influence of the materialist worldview, we find ugliness. It's nowhere more apparent than in the hideous, depersonalizing art of Stalinist Russia and the modern-art rooms of most public museums, which aren't exactly bastions of capitalism.

Myth no. 7: The Artsy Myth (confusing aesthetic judgments with economic arguments)

The cultural critique of capitalism reduces to little more than aesthetics masquerading as economics. Good tastes don't suspend the truths of economics. Moreover, it is the materialist worldview that even denies the divine, and not capitalism, that is the greater driver of ugliness in modern culture. By pursuing the wrong villain, Christians squander their energies, which could be better spent battling the real culprit—an insidious worldview that denies the great truths of Christianity, truths that led to the greatest flowering of art and architecture, both high and common, that the world has ever known.

Eight

Are We Going to Use Up All the Resources?

You've heard it a million times. The earth is overpopulated. We're breeding like rabbits and eating like locusts, and soon we'll run out of food, farmland, and fuel. We are members of the crew of "Spaceship Earth." We have to preserve our dwindling supply of provisions or our mission will soon be aborted. Our industrial technology is poisoning the water, the soil, and the air. If we don't make radical changes now, it will be too late. We'll destroy the earth. We'll all die.

Fears about running out of resources are as old as the human race. But this planet-sized vision of disaster started with the nineteenth-century demographer Thomas Malthus. In his early writings, he predicted that a swelling human population would quickly overtake food production and lead to widespread famine. He was wrong, of course: it didn't happen and still hasn't. He changed his tune in his later years; but that hasn't discouraged a long line of Malthus wannabes from updating his old argument by moving the date forward, like those Bible-prophecy experts who keep missing and then moving the date of Jesus's return.

The 1960s and '70s were dense with doomsday declarations. Outfits like the Club of Rome warned that we were using up all our resources. And in *The Population Bomb* (1968), biologist Paul Ehrlich wrote that England had just a 50 percent chance of making it to the end of the twentieth century.[1] His book opened with mathematical certainty: "The battle to feed all of humanity is over. In the 1970s the world will undergo famines—hundreds of millions of people are going to starve to death."[2] Despite their

perfect losing streak, such warnings remain fodder for practically every fund-raising letter ever sent out by Greenpeace or the
Foundation for Deep Ecology.

Regrettably, these claims aren't limited to secular environmentalists. Prominent Christians say the same thing. For instance, in
1997, Ron Sider made this assertion:

> Economic life today, especially in industrialized societies, is
> producing such severe environmental pollution and degra-
> dation that the future for everyone—rich and poor alike—
> is endangered. We are destroying our air, forests, lands,
> and water so rapidly that we face disastrous problems in
> the next century unless we make major changes. . . . We
> overfish our seas, pollute our atmosphere, exhaust our sup-
> plies of fresh water, and destroy precious topsoil, forests,
> and unique species lovingly shaped by the Creator.[3]

Sider, like others, blames this wanton rape and pillage on "present economic patterns" by which we "maintain and expand our
material abundance."[4]

If Sider and the rest of the choir are right, modern capitalism
is just a Ponzi scheme where we rob from future generations by
using up all the limited resources now. Our economic growth,
they claim, is unsustainable. We reap the benefits; our grandchil-
dren, the costs. Most people accept this claim without question.
Whenever I speak about environmental issues, I always get asked
about the "fact" that we're depleting the earth's resources. But
it's not a fact. The truth, despite untutored common sense, is just
the opposite. As long as we can preserve our economic freedom
and the spirit of enterprise, we will not use up all our resources,
nor will we run out of food, water, or energy. The prophets of
doom are demonstrably wrong.

Well, they're mostly wrong. It's the smidgeon of truth in their
view that gives it credibility. Yes, Earth is a planet. It has a finite
volume, mass, and surface area. Its surface area is 510,065,600
square kilometers—148,939,100 square kilometers of land. We
can reach only the tiny bit of Earth's mass near its surface. We

draw resources like oil and coal from that tiny surface to fuel our economy. There's only so much of the stuff to go around. Whatever we use now, future generations will lack. In the short run, we often experience a scarcity of at least some natural resources. Otherwise, they'd all be cheap as bad dirt. Given these truths, it seems obvious that we are quickly depleting limited and irreplaceable natural resources. And it *is* obvious, until you understand the full meaning of "resource" and the wonder of human creativity.

STEWARDS

Debates about resources and the environment have a way a producing more heat than light. Whenever I speak to large crowds about this subject, there are usually a few people who storm out before I'm finished. They seem to assume that, because I criticize certain aspects of environmentalism, I hate the environment. So let's remind ourselves of the truths that Christians all agree on. First, this is God's world, not ours: "The Earth is the Lord's and all that is in it, the world, and those who live in it" (Ps. 24:1). Although we may own legal property, God holds the ultimate title on everything.

Second, when God created human beings as his image bearers, he commanded us: "Be fruitful and multiply, and fill the earth and subdue it; have dominion over the fish of the sea and over the birds of the air and over every living thing that moves upon the earth." This command baffles some readers. In 1967, Lynn White wrote a famous paper in the journal *Science* blaming this biblical view of man's dominion for environmental problems: "By destroying pagan animism, Christianity made it possible to exploit nature. . . . We shall continue to have a worsening ecologic crisis until we reject the Christian axiom that nature has no reason for existence save to serve man."[5] Almost every student in environmental ethics learns about White's argument. But they're rarely told that White misread the text. The Bible nowhere says that "nature has no reason for existence save to serve man." And Genesis 1 speaks of "dominion," not "domination." The Bible is describing the rule of benevolent stewards who represent the good God to the rest of his creation. As stewards, we're responsible for how we treat and use it. We're part of God's good creation, as well as its crowning achievement. "God saw everything that he had made," says Genesis, "and indeed, it was very good" (1:31).

Third, God intends for us to use and transform the natural world around us for good purposes. Adam and Eve were put in a garden before the fall, and told to "tend and watch over it" (Gen. 2:15). Working and transforming the earth is part of God's blessing, not a curse. Work and change did not arise from our fall into sin. The fall simply turned work into toil, since the ground would resist our efforts to cultivate it.

Fourth, the world is good but it—all of it—is now fallen. "The creation was subjected to futility," Paul says, "the whole creation has been groaning in labor pains until now" (Rom. 8:20, 22). As fallen creatures in a fallen world, then, it's no surprise that we can mess things up. We can and do pollute. We can and do act irresponsibly, ignoring the unintended but bad consequences of our actions.

Fifth, we can't build God's kingdom on our own, but our actions can make a difference for good. As Christ foreshadows the kingdom to come, so, too, does the work of his body—the church. Caring for God's creation, or at least the tiny portion we can affect, is one of our responsibilities.

Almost all Christians can agree on these five basic principles. But we often fail to work them out logically. As a result, we end up adopting the materialistic assumptions of our culture.

"RESOURCES" AND PRICE

When we hear "resource," we think of stuff you can weigh or count: oil in the ground, land underfoot, water in a lake or aquifer, gold bars in a bank vault. Some resources are renewable: as long as we don't cut down more trees than we plant, for instance, there's no reason to worry that we'll run out of lumber any time soon. And aquifers tend to fill back up as long as we don't suck them dry too quickly. Other resources aren't renewable: oil and coal, for instance. So far as we know, oil reservoirs don't refill. That's why we're always hearing dire warnings that we're past the peak of oil reserves and that they're now dwindling.

On the surface, the warning makes sense. But only on the surface. The problem with these warnings is that they are almost always based on proven or *known* oil reserves. How much oil we know about doesn't tell us how much oil exists. Discovering an oil reserve costs money. BP or ExxonMobil or Arco has to lay out a lot of cash digging dry holes before it discovers a new reserve. That's a crucial point. "How much of any given natural resource is known to exist depends on how much it costs to know," says economist Thomas Sowell.[6]

As the current supply dwindles, or as demand spikes, the price per barrel goes up. At some point, the price gets high enough that it encourages oil companies to seek out new reserves even in costlier locations (since they can make a profit at the new, higher price). When they find a new reserve, they still have to tap it, estimate its size, transport it, refine it, and deliver it. In this way, new supplies of oil flood the market and again regulate the price. In recent years, jumps in oil prices have set in motion precisely this process, making profitable forms of exploration that were too expensive before.

Known reserves tell us how much it's worth to know about right now, not how much total oil there is to discover or exploit. We can be confident that we're nowhere near running out of oil simply because oil companies aren't hoarding oil and the price of gasoline isn't a million dollars a gallon.

OK. But if there's a fixed supply of oil, won't we eventually run out of it if we keep burning it like we are now? Yes, but the mischief is in the word *if*. Let's assume we're still using oil for energy in a hundred years and it becomes so costly to tap new reserves that the next barrel is going to cost one thousand dollars (in today's dollars), with economists further predicting that oil will soon go to ten thousand dollars a barrel. Long before it got to ten thousand dollars a barrel, though, the high price would signal to everyone that it was time to carpool, take the bus, hitchhike, or switch to a cheaper source of energy. That's what prices *do*. And they do it far better than any nanny-state regulation. This isn't happening now because, for most uses, oil is still the best and cheapest source of energy available.

CREATING RESOURCES

In 1980, economist Julian Simon made a famous bet with biologist and environmentalist doomsayer Paul Ehrlich. Ehrlich and other environmentalists had been predicting for years that we were about to use up the earth's natural resources. Simon realized that if that were true, the law of supply and demand would force up the prices of the resources as they got scarcer and scarcer. So Simon did the unthinkable. He publicly bet one thousand dollars that over the next ten years, the real prices of any five commodities would go *down,* not up. Ehrlich and his team took up the challenge and picked five commodities: nickel, tin, tungsten, chromium, and copper.

Ehrlich's team lost the bet. The real price of all five commodities went *down* in the 1980s. Of course, when demand increases faster than supply, a price will go up. And that's what sometimes happens in commodities markets. Ehrlich might have gotten lucky. Demand might have outstripped supply in the '80s. But Simon understood what Ehrlich did not—that the amount of stuff hiding in the ground somewhere is far less important than human beings devising new ways to access and exploit the stuff. Because we develop new ways of exploring, mining, and refining, future resources are often cheaper to acquire than current resources, even under pressures of greater demand.[7] Simon was banking on that fact.

Simon also knew another indisputable fact: over time, virtually any natural resource you can think of—oil, copper, mercury, coal, whatever, *has gotten less scarce, more plentiful, and therefore, less expensive.* This is easily established by looking at the price trends of resources historically. Adjusting for inflation and over the long run, they always go down, not up. Stop and think about that for a minute, because it probably sounds impossible.

But it's not. It's a well-established fact.[8] It just contradicts our expectations because we tend to notice only short-term scarcities and spikes in the cost of, say, a gallon of gas. We don't notice the long-term trends.

When facts contradict expectations, we should reexamine our expectations. What falsehood are we assuming that makes us expect the wrong thing? Remember the zero-sum game and materialist myths? They make spectacular encore appearances in debates about resources, and almost everyone stays to watch. That's probably because we're thinking of the material part of a resource. We fear that we're running out of resources because we're thinking of them merely as some finite amount of physical stuff. That seems like common sense; but it's wrong. Resources aren't just there in a tank or in the ground. On the contrary. We *create* resources.

Sound crazy? Think about it. For centuries, oil was viewed mainly as an irritating pollutant. No one even knew about all the oil deposits thousands of feet under Texas ranches and five or six feet under the Arabian deserts. When people did find some of it bubbling to the surface, they were none too pleased. But in the 1840s, someone figured out how to refine oil into kerosene, and someone else figured out that kerosene was useful in lamps and furnaces. Whales were pleased with these developments, since kerosene replaced whale oil.

Suddenly, there was a demand for petroleum. In 1859 the first oil well was dug (sixty-nine feet down) by a guy named Edwin Drake in western Pennsylvania. It rendered fifteen barrels a day. Soon other people were finding oil deposits in all sorts of places.

Black gold really took off with the invention of cars and the internal combustion engine. Since then, we have devised all sorts of ways to explore, refine, and use it more and more efficiently. Think how different oil is now than when it was gurgling up and making a mess in Mr. McIlhaney's cornfield. The differences lie not so much in the nature of oil but in the vision and ingenuity of man.

Most resources are resources only because human beings are involved in some way. In fact, over time the matter in a material resource matters less than how human beings creatively transform it for some use: wood is transformed into fuel and lumber, straw into baskets, clay into pots and bricks, fur into coats, fields into farms, manure into fertilizer, oil into gasoline and kerosene, stones into walls, iron ore into spearheads, cotton into clothing, copper into phone lines, sand into computer chips and fiber-optic cables, light into lasers.

That's only part of the story. We don't just figure out how to use resources more efficiently. We discover, and create, fundamentally different types of resources. At every stage, some doomster can do a little math and predict that the current resource will soon be depleted. And he will almost be right. In fact, people in every era of recorded history have worried about running out of whatever resource they're using at the time. England began to experience lumber shortages in the 1600s. They got so severe in the 1700s that the island came close to being stripped of its forests. People feared a complete loss of wood. So what happened? Wood became too costly to use as fuel in most places. That encouraged innovation with other resources, like coal. The English eventually switched to coal, and over time, English forests returned.

The process was hardly inevitable. It involved all manner of effort and ingenuity, usually brought on by rising scarcity, which led to rising prices. Because of the role of prices, scarcity and creativity conspire to get us to the next level, to the next resource or the next technological breakthrough. Necessity is indeed the mother of invention; but a human creator is the father, unless the creator is a woman, in which case necessity is the father of invention and . . . well, you get the idea.

So after the switch to coal, did all of England rest easy and quit worrying about running out of resources? Hardly. In 1865, a prominent social scientist named W. Stanley Jevons wrote a book proving to his satisfaction that England would soon exhaust its coal, and the economy would grind to a halt. It didn't happen, and there's still plenty of coal available more than 140 years later.

Did such experiences teach the doomsayers to qualify their warnings? Nope. The same unqualified claims of disaster quickly emerged with petroleum as well and have continued down to the present, despite one prediction after another biting the oil-stained dust.[9]

In fact, if we get past the short-term spikes in energy prices and look at the big historical picture, it quickly becomes clear that energy has grown cheaper and cheaper over time. When England switched from wood to coal, energy became *more abundant* and energy costs went *down*. Over the long run, energy has continued to become less scarce and less costly. Think about it. You can get free electricity for your laptop from random outlets in most American airports. Such a thing would have been unimaginable in earlier times, as it still is in impoverished regions today. If you're focused on the resource we happen to be using right now, it's easy to miss the fact that energy costs go down over time.

> "The Stone Age came to an end not for a lack of stones, and the oil age will end, but not for a lack of oil."
> —Sheik Yamani,
> Saudi Arabian oil minister
> and founder of OPEC

History again and again teaches a basic lesson: the fact that there's a fixed supply of wood or coal or oil or uranium doesn't mean that we are doomed to run out of energy supplies. The image conjured up is of some fixed pot of stuff called "energy," where the big kids are getting more than their fair share. We need to use less so that others can have more. So complains feminist theologian Sallie McFague: "We do not love nature or care for two-thirds of the world's people if we who are 20% of the population use more than 80% of the world's energy. There is not enough energy on the planet for all people to live as we do (and increasingly, most want to) or for the planet to remain in working order if all try to live this way."[10]

But what exactly does McFague mean by "80% of the world's energy"? Presumably the "world" refers to the earth, and not the universe. But she can't be thinking of all the energy in all the earth's matter. That amount of energy is almost unimaginably high. At any time, we're using only the tiniest fraction of that. Besides, the earth isn't a closed system. We get energy from the sun. Thus, she can't be talking about all the currently feasible energy sources, since there's a whole lot of uranium, sunlight, wind, waves, and river currents, for instance, that we're not using for energy. And she's quantified the total so precisely that she can't be referring to all the oil, since we don't know how much oil the earth has.

So she must be referring to the total amount of available energy being *produced* at the moment. And that one little verb changes the picture entirely, since it begs the question, Who's producing it? Usable energy isn't just sitting in a battery some-where, first-come, first-served. Somebody has to produce it. So unless the energy consumers are stealing from the energy pro-ducers, what's the problem? Some places produce, buy, and con-sume more energy than other places. The problem isn't that some places are able to produce or buy ample energy. The problem is that other places are not.

McFague may have a wonderful imagination in other areas of life, but concerning energy as a resource, not so much. To be fair, this kind of zero-sum thinking is easy for anyone to fall into. We're like the circle protagonists in the two-dimensional Flatland world of author Edwin Abbott. The circle inhabitants there experience spheres passing through Flatland as if they were other circles.[11] The circles can't imagine a three-dimensional circle called a "sphere." Like those circles living in Flatland, we're tempted to think of resources strictly from within the framework of our tiny slice of time and space.

Myth no. 8: The Freeze-Frame Myth (believing that things always stay the same—for example, assuming that population trends will continue indefinitely, or treating a current "natural resource" as if it will always be needed)

Instead, we need to think not in two or even three dimensions but in five dimensions, taking into account not only the three space dimensions, but also the dimensions of time and human creativity. We need to reflect on how things change over time, and how, in a free economy, over time, human knowledge and creativity build upon themselves. Over time, we substitute more and more of the matter in a resource with the unique resource of mind called *information*. As far as the creation of wealth and resources goes, that is the most important dimension of all.

If we keep the *long-term* past trends in mind rather than just our little slice of time, we should expect scarcity and creativity to conspire in the future, as they have in the past. Of course, looking forward, we can only guess at the resources that will replace oil once it becomes too expensive. They may be something we can easily imagine, like fission reactors using uranium rods. Despite bad press, nuclear energy is much cleaner and safer than the alternatives. Even France now gets over 70 percent of its energy from nuclear power plants. Or the future's main energy source may be fusion reactors using deuterium culled from seawater, harnessing a type of atomic reaction even more energy-rich than fission. Or it may be something more exotic but right around the corner, like paint that you can put on anything that captures usable energy from the sun. Who knows? The point is that, given what we know historically about how prices and inventors work in a free economy, we have far more reason to expect a solution than a disaster.

YES, BUT . . .

Of course, if you've watched the news lately, you're probably thinking: OK, maybe we won't run out of every resource, but aren't we messing things up with the resources we're using now? What about the carbon dioxide and other greenhouse gases we're pumping into the atmosphere? Isn't our energy use leading to a planet that is less and less habitable?

That's certainly the official story of the mainstream media. We're surrounded by it. In April 2007, I went to speak at Southern Methodist University. Earth Day was coming up, and a couple of days before, musician Sheryl Crow had performed on campus as part of her Global Warming tour. As I waited for my flight at the airport, within five square feet of wall space on a magazine stand there was *Newsweek* with Arnold Schwarzenegger on the cover: "Save the Planet—Or Else," it read; *Vanity Fair's* second annual "Green Issue," sported a scowling Leonardo diCaprio on a glacier, outfitted with crampons and standing beside a baby polar pear; *Elle's* "Green Issue," featuring the story "Eco-chics Heat Up"; *Fortune's* "Green Issue," with radical environmentalist Yvon Chouinard, founder of Patagonia, smiling on the cover. And there were plenty of secondary eco-stories in other magazines. Al Gore had recently won an Academy Award for his documentary, *An Inconvenient Truth.* The international Live Earth concerts would soon be televised around the world with Gore's help. And Gore would win a Nobel Peace Prize (along with the U.N.'s Intergovernmental Panel on Climate Change) for his environmental advocacy. This was all in the space of a few months. I now receive a dozen news stories every day telling me with unqualified certainty that we're heating up the planet and it's going to be a disaster unless we let a U.N. committee take over our energy policy.

And you thought the peer pressure on the junior high playground was bad.

With all of these voices telling us the sky is falling, almost no one wants to be left out. Plenty of Christians are rushing to catch up: finally, here's a chance to be hip, relevant, and respected by the wider culture and mainstream media without having to go along with something like abortion on demand or gay marriage. The Evangelical Environmental Network sponsored the What Would Jesus Drive? campaign in 2002, and the mainstream media loved it. Ditto with the 2006 Evangelical Climate Initiative, which featured eighty-six evangelical leaders calling for the federal government to restrict carbon dioxide emissions.

Is it time to climb on board? Certainly any thoughtful Christian should take a hard look at the evidence for human-induced global warming. But in doing so, it's also a good idea to keep your skeptic's hat close at hand. This global-warming frenzy has all the marks of a classic sociological phenomenon known as *groupthink*.[12] As a rule, when partisans appeal to "consensus" in a scientific dispute, chances are that: (a) there's not a consensus; (b) the partisans are trying to silence dissent and marginalize dissenters; and (c) the evidence for their view isn't that great. No scientist worth her salt would appeal to consensus in making a case for, say, the theory of continental drift (plate tectonics; and here there really is a rough consensus). The scientist doesn't need to. She can simply lay out the evidence. But when it comes to predictions about what's going to happen fifty years from now, based on highly speculative computer models that every expert knows do not accurately represent reality, suddenly there's talk of a uniform consensus.

The ad hominem arguments are another red flag.[13] Challenge any part of the official story publicly, and you won't be met with persuasive arguments and indisputable facts. You'll be compared with apologists for mass murderers: "I would like to say we're at a point where global warming is impossible to deny," wrote columnist-turned-climate-scientist Ellen Goodman in the *Boston Globe*. "Let's just say that global warming deniers are now on a par with Holocaust deniers."[14]

Right. None of this proves that we're not causing global warming. But these are the tried-and-true tactics for protecting an intellectual orthodoxy, not for making a credible scientific argument. Every age has at least one such orthodoxy. It's easy to spot past ones once the pressure's off: At one time most of the officially smart people believed in Martians, Marxism, eugenics, and global cooling. We look at those old orthodoxies now and wonder how so many people could have been so wrong. The challenge is to discern the present wrongheaded orthodoxy when the social pressure to conform is intense.

Regrettably, the environmental movement, which started with many noble goals, has now become the last line of defense in the left's attempt to stem the tide of global capitalism. Like communism, the environmental movement provides an all-encompassing vision of reality. Unlike communism, it has the benefits of a gratifying nature spirituality. When Al Gore accepted the 2007 Nobel Prize (shared with the IPCC), for instance, he explained: "The climate crisis is not a political issue, it is a moral and spiritual challenge to all of humanity. It is also our greatest opportunity to lift global consciousness to a higher level."[15] He also has called for environmental protection to become our "central organizing principle." No mere scientific hypothesis here. It's impossible to understand the modern environmental movement without taking these religious dimensions into account.

> "We have to offer up scary scenarios, make simplified, dramatic statements, and make little mention of any doubts we have. Each of us has to decide what the right balance is between being effective and being honest."
> —National Center for Atmospheric Research (NCAR) climate researcher and global-warming action promoter Steven Schneider

Unfortunately, the global-warming scare is now the left's best chance to establish more national (or international) control of the economy, not in the name of efficiency or economic justice, but in the name of the earth and future generations. In a recent documentary, Greenpeace founder Patrick Moore said that by the mid-1980s, environmentalists had succeeded in their primary goals.[16] So the activists got more extreme. Then, with the end of the Cold War, many left-wing activists joined the environmental movement without changing their philosophy. Environmentalism then became, as Moore put it, the "new guise for anti-capitalism."[17] If you don't believe me, just listen to the rhetoric and the proposed solutions. They almost always involve more government (or U.N.) control and less economic freedom, and treat economic progress as the problem, not the solution.

GLOBAL WARMING: A FEW QUESTIONS

The topic of global warming is complicated. I can't possibly do it justice here. But the basic outline is easy to understand. The central claim about global warming is that human beings, by releasing carbon dioxide and other greenhouse gases into the atmosphere, are creating catastrophic climate change. And if we don't do something about it soon, it will be too late. However you judge it, the claim itself bundles together issues that need to be considered one at a time. There are at least four questions you can ask about global warming:

1. Is the planet warming?
2. If the planet is warming, is human activity (like carbon dioxide emissions) causing it?
3. If the planet is warming, and we're causing it, is that bad overall?
4. If the planet is warming, we're causing it, and that's bad, would the policies commonly advocated (e.g., the Kyoto Protocol, legislative restrictions on carbon dioxide emissions) make any difference?

Based on current evidence, to question one I would answer: "Probably." That is, we're probably in a slight warming trend, if you pick a popular but arbitrary baseline of, say, 1870. Despite what you've been told, this is the only question on which there really is a scientific consensus. There's plenty of debate and no consensus on the other stuff.[18]

Incidentally, we're actually cooler now than in the year 1000, so which baseline you pick makes a big difference. In a trivial sense, though, the climate is "changing"—the earth's average temperature is increasing right now. I say this is trivial, because we know from natural "data recorders" like ice cores that the earth's climate is *always* changing. In fact, the last several thousand years of recorded human history have been strangely mild. The changes we are currently experiencing are well within the known natural variations in global climate.[19]

What about question two? Are carbon dioxide emissions causing this warming? The question isn't whether carbon dioxide is a greenhouse gas. The question is whether the carbon dioxide we've put into the atmosphere in the last century is the primary cause of the warming. After all, one of the many natural feedback mechanisms could be mitigating its effects. For example, in some cases an increase in carbon dioxide leads to more plant growth, which in turn sequesters the carbon dioxide. This is one of many examples of a natural feedback process that makes it unimaginably hard to make predictions about the future climate.

Add to that all the other possible causes and contributors, like changes in the energy output and magnetic activity from the sun. Recent data suggest that it's also gotten warmer on Mars.[20] ExxonMobil, Texaco, and their oil buddies surely didn't cause that. With predictions of future global warming, almost all the work is done by the assumptions plugged into the computer models, not by direct evidence of what's causing warming. That's why, at the moment, the prudential answer to question two would be: "We don't know."[21]

What about question three? Is it obvious that global warming would be bad overall? No, it's not. It might lead to droughts in some places, but to warmer, wetter, more productive weather elsewhere. The total might be a net gain. We already know that, within limits, warm weather is better than cold weather—that's why most Canadians live close to their southern border with the United States. (It's not because they love Americans.) And less energy is used in warm weather than in cold. So we have little reason to assume that some warming would be bad.

Recent estimates point in that direction. Bjørn Lomborg calculates that about two hundred thousand people die in Europe each year from extreme heat but that 1.5 million die from extreme cold.[22] You can't focus just on the two hundred thousand, as is the custom in almost every story about the bad effects of warming. You have to weigh those deaths against the 1.5 million. Another study suggests that human carbon dioxide emissions could be preventing an overdue ice age.[23] Neither of these proves warming would be better than cooling on balance. But these are just the sorts of factors that any serious judgment on the issue will have to take into account.

In any case, we don't know what the optimum average global temperature is, so for all we know, warming is good rather than bad.

Finally, remember the economic trade-off. According to an article in *National Review,* "Earth got about 0.7 degrees Celsius warmer in the twentieth century while it increased its GDP by 1,800 percent, by one estimate."[24] Even if we caused all the warming, that's a great bargain.

What about question four? Is it obvious that reducing carbon dioxide emissions in the United States, for example, would make much difference? No, it's not. Take the Kyoto Protocol, which requires participating countries to reduce annual emissions to 5.2 percent below 1990 levels. The official estimate is that such a reduction would slow current warming by an undetectable 0.07 degrees centigrade by 2050. In other words, it wouldn't do squat, but it would cost a lot. To comply, the estimated cost to the worldwide economy would be between $10 trillion and $50 trillion. Imagine what it would cost to reduce carbon emissions by 80 to 90 percent without benefit of a new form of energy. In contrast, the economists who form the "Copenhagen Consensus" have estimated that it would cost about $200 billion to outfit the rest of the world with water-sanitation capacity, which is 50 to 250 times cheaper than the estimated cost of Kyoto. Unless we're interested in practicing random acts of piety that don't do anything except squander money that would be much better spent elsewhere, we should be skeptical of the Kyoto Protocol and other similar attempts to restrict carbon emissions by fiat.[25]

IT'S GETTING BETTER, NOT WORSE

It's only by turning carbon dioxide (a.k.a. plant food) into a pollutant that we've missed all the good news about *long-term* environmental improvement in modern societies. On almost every measure, we are healthier than we have ever been, and our environment is cleaner than it has been even in the recent past.[26] Over time, we use more-efficient and less environmentally destructive forms of energy. In the developed world, most of the really important measures have shown enormous improvement in recent decades, not worsening: wealth, infant mortality, life expectancy, nutrition, and the leading environmental indicators such as air and water quality, soil erosion, and toxic releases.[27] In general, the wealthier a country is, the more environmentally sustainable it is.[28]

We've long since solved and so forgotten about the most devastating environmental problems that still plague the poorest parts of the world. They're the ones caused by bacteria, viruses, insects, and particulate matter. How many people do you know who have died of the plague, malaria, food or water poisoning, smallpox, cholera, typhus, dengue fever, or any of the other environmental killers? How many people do you know with respiratory illness from proximity to burning wood or dung in poorly ventilated houses? Probably few or none, unless you know missionaries in Africa or Asia. Instead, we complain about mysterious chemicals in our food that kill no one, and fret about the clean water that comes out of every tap in the United States because it doesn't taste as good as bottled water from a well in France or Fiji.

Long-term trends in life expectancy—surely an important indicator of environmental health—are good, not bad. Those trends are the result of human innovations made possible by societies that enjoy political and economic freedom. Life expectancy has gone up worldwide in the last fifty years, even in poor countries. The trends go down only in countries with widespread war and extremely corrupt and despotic governments.[29]

Before listing its litany of traditional complaints, even the United Nations admitted as much. Its unreported 2007 document titled "State of the Future" began: "People around the world are becoming healthier, wealthier, better educated, more peaceful, more connected, and they are living longer."[30] The document even goes so far as to admit that these improvements are the fruit of free trade and technology.

Of course, the fact that things are getting better doesn't mean the environment is as good as it can be. We *should* continue to seek solutions to real, well-known, tangible pollution problems, especially at the local level. Sometimes we need laws that protect the "commons," those places owned by the government or no one in particular. Aristotle understood twenty-four hundred years ago that people treat their own property better than they treat other people's property or public land. We tend to act less responsibly when we are not directly affected by our actions. We're more likely to keep our own bathroom clean than to keep the airport bathroom clean. Because of this tendency, strong private-property laws are often the best ways to encourage people to act in environmentally friendly ways.

Have you ever wondered why we almost wiped out the buffalo during the western expansion of the United States but cattle have never been at risk of extinction? The only difference is that ranchers own cattle. When the West was being settled, no one owned the buffalo, so those who hunted them rarely bore the full cost of their actions. Now that they live on ranches, now that they are owned by ranchers, buffalo are beginning to recover.

But as with resources, so with pollution. It's easy to work yourself into a tizzy by extrapolating some recent pollution trend. A New Yorker in the mid-1800s could have fretted about the imminent demise of streets, which surely would eventually be filled neck deep with horse poop from all the horse-drawn carriages. He would never have imagined that a hundred years later people would get around in manure-free air-conditioned metal boxes, or that the only horses on New York streets would carry specialized policemen and newlyweds. Most resources aren't merely matter. They emerge where man and matter meet. And just as human ingenuity leads to new resources in a market economy, so, too, it can forge solutions to real environmental problems.

"One of the main figures of the environmental movement is to pass off a temporary truism as an important
indicator of decline."
                                        —Bjørn Lomborg

Remember, environmental protection is a costly good. Societies start to worry about the environment once they have
solved basic problems of survival. Americans with four-bedroom
houses, three square meals, two cars, and one dog are much
more likely to fret about recycling, topsoil erosion, and the plight
of the fish in the local reservoir than are Africans who live in
shantytowns. The developing world will become more environmentally conscious if and when it becomes economically wealthier. As Warren Brookes has said: "The learning curve is green."

Moreover, the wealthier you are, the easier it is to adapt to
change. Any change in the economy or the climate is going to
hit the poor the hardest because they have the fewest means
to *adapt*. So even if the current warming trend continues and
makes it harder on the poor, it's much wiser for us to help them
to become wealthier rather than to stage quixotic campaigns to
regulate civilization back to the Stone Age.

Whatever your opinion about the environment, it should be
based on real data, not *Dateline NBC*. You're always free to fear
the future. But remember: every predicted global environmental
catastrophe based on current trends has proved false.[31] If we
look at long-term historical trends, in contrast, the evidence of
declining energy costs, increasing energy abundance, and growing prosperity provides no basis for such pessimism.

WHAT IS MAN?

There's a chilling scene in the 1999 film *The Matrix*. The story's
backdrop is a near future where intelligent machines take over
civilization and use comatose-induced human beings as their
energy source. To keep the humans docile, the machines create
an illusory computer-projected world, the Matrix, so the humans
will believe everything is business as usual, even though they're

all really encased underground in liquid vats. The story revolves around a group of rebels who have escaped from the Matrix. At one point, the arch-villain, Agent Smith (a sort of personified computer program), speaks to Morpheus, one of the main freedom fighters, who is bound to a chair and hooked up to electroshock equipment:

> I'd like to share a revelation that I've had during my time here. It came to me when I tried to classify your species. I realized that you're not actually mammals. Every mammal on this planet instinctively develops a natural equilibrium with the surrounding environment, but you humans do not. You move to an area, and you multiply, and multiply, until every natural resource is consumed. The only way you can survive is to spread to another area. There is another organism on this planet that follows the same pattern. A virus. Human beings are a disease, a cancer of this planet, you are a plague, and we are the cure.

This may sound like a crazy sci-fi plot, but Agent Smith's attitude is frighteningly common.

Certain ideas are like geese: they tend to travel together. The idea that we're destroying the planet and using up all our resources usually keeps company with another fashionable idea: that the earth is overpopulated—with people, that is. The logic seems airtight: the more people there are, the more Big Macs and Krispy Kremes you need to feed them. To get Big Macs, you need cows and farms and a little lettuce and lots of Thousand Island dressing and fertilizer and bakeries and trucks and diesel fuel to deliver them to McDonalds. Multiply by 7 billion, and it gets easy to worry. Just how many Big Macs, or guys named Big Mac who eat too many Big Macs, can the earth hold?

The outcome of such ideas ranges from the macabre to the monstrous. The macabre includes the Voluntary Human Extinction Movement (VHEMT). As they put it: "Phasing out the human race by voluntarily ceasing to breed will allow Earth's biosphere to return to good health. Crowded conditions and resource shortages will improve as we become less dense."[32] (At least they'd prefer you cease breeding voluntarily.) This childbearing-is-bad posture has the added bonus of gratifying the ego. Now the decision to skip the hard work of bearing and raising children can be viewed as sacrificial, as a moral virtue, rather than as what it usually is—a desire to skip the expense and challenges of parenting in favor of the sort of lifestyles you find on TV sitcoms like *Friends*.

At the monstrous extreme is what the Reverend Robert Sirico calls "humanophobia": the belief that life and the earth would be a lot better off if most of us were, well, dead. Sirico, of course, rejects this view. One might suppose that every respectable member of society would reject such a view. Unfortunately, that isn't the case. In a speech to the Texas Academy of Science in March 2006, University of Texas professor Eric Pianka gained unwanted publicity by waxing longingly for a mutated Ebola virus that would wipe out 90 percent of the human population. You might think Pianka was just one eccentric scientist who said something crazy. But he received a standing ovation from hundreds of members of the Texas Academy of Science.[33]

His views are common among environmentalists. A couple of years ago, I received a letter from the retired curator of the Field Museum of Natural History in Chicago, who shared Pianka's sentiments. He said,

> Still, adding over seventy million new humans to the planet each year, the future looks pretty bleak to me. Surely, the Black Death was one of the best things that ever happened to Europe: elevating the worth of human labor, reducing environmental degradation, and, rather promptly, producing the Renaissance. From where I sit, Planet Earth could use another major human pandemic, and pronto!

This kind of thinking is well on its way to becoming respectable opinion. You can probably drop bombs like this at fashionable Manhattan cocktail parties even now and raise nary an eyebrow. And it's quickly leaking into the wider culture. A recent story from the United Kingdom reported on environmental activists who have abortions and sterilize themselves to save the planet from the scourge of humanity. "Having children is selfish. It's all about maintaining your genetic line at the expense of the planet," said one such activist. "Every person who is born uses more food, more water, more land, more fossil fuels, more trees and produces more rubbish, more pollution, more greenhouse gases, and adds to the problem of over-population."[34]

There are several problems with this line of thinking. First is the way it pits us against nature. Many environmentalists accept antihuman assumptions without thinking about it. They treat human beings as alien parasites separate from nature, even define nature without reference to us. For instance, in a *New York Times* editorial complaining about how much of the planet human beings have affected, Verlyn Klinkenborg asserted without argument, "The essence of nature is that it is not 'for people.'"[35] That's not scientific evidence. It's a highly debatable theological assertion.

Second, it assumes that we mindlessly breed like bunnies and bacteria until we starve to death. We don't. Historically, population growth has adjusted for all sorts of reasons. In recent decades, in those parts of the world that are wealthier and living longer, population growth rates have *decreased*.[36] Most of the wealthiest parts of the world now barely reproduce at the replacement rate of 2.1 children per couple. And much of Europe is now reproducing *below* replacement rate. Walk the streets of Rome and Copenhagen, and you'll be struck by how few babies you see.

Other parts of the world are at earlier stages of industrialization and have higher fertility rates. As they grow wealthier, however, their fertility rates probably will decline as well. Such trends have led even the population-control partisans at the United Nations to predict that human population will level off at about 9 billion in 2050. (Right now, it's about 6.6 billion).[37]

These are the demographic facts. I don't concede the claim that we're in danger of an unsustainable population explosion. And I don't concede the more insidious assumption behind population control: that human beings mainly consume resources like gluttonous swarms of locusts. Too many environmentalists treat this as a self-evident truth.

In some situations, people *are* mainly consumers. Babies are mainly consumers. (Shame on them!) But most people in free societies grow up to produce more resources than they consume. This is the Achilles' heel of the misanthropic strain of modern environmentalism: free societies allow human beings to be fruitful and multiply rather than merely consume. If they didn't, market economies would shrink. Instead, over the long run, they grow.

In free markets characterized by the rule of law and limited government, output per capita goes up, which means that the productivity of our labor increases. Our labor is enhanced by "labor-saving devices." That's why Spanish Scholastic philosopher Luis Molina referred to human productivity as the "fruit of our ingenuity."

We know market economies grow. So why do we often fall for claims that contradict what we already know? Because we forget what late economist Julian Simon called "the ultimate resource"—the creative imagination of human beings living in a free society. The more human beings in free societies there are, the more inventors, producers, problem solvers, and creators there are to transform material resources and to create new resources. Man, not matter, is the ultimate resource.[38]

> "If you want 1 year of prosperity, grow grain. If you want 10 years of prosperity, grow trees. If you want 100 years of prosperity, grow people."
> —Chinese proverb

This is the most important economic truth, and Christians should have expected it all along. It's ironic that a nonreligious economist like Julian Simon would see that truth so clearly, while so many of our Christian leaders miss it.

One Christian leader who didn't miss it was Pope John Paul II. In his 1991 encyclical, *Centesimus Annus,* he said, "Indeed, besides the earth, man's principal resource is *man himself.* His intelligence enables him to discover the earth's productive potential and the many different ways in which human needs can be satisfied."[39] Read that again: "Man's principle resource is *man himself.*" Grasp that, and you'll know why we're not going to run out of resources.

Does such a stance demand hope? Yes, but a hope based on reason and experience, a hope based on the revealed truth that we are creators made in the image of *the* Creator. And hope, remember, is one of the three theological virtues. Humanophobia driven by despair is not.

Working All Things Together for Good

If I've conveyed anything in the previous pages, I hope it's the oft-neglected truth that the creation of wealth has as much to do with spirit as with matter. Christians should be the first to understand this, since we know that human beings are a unique hybrid of the spiritual and the material.

We've looked at the problem of extreme poverty, and how many of our best-laid plans fail to solve it. We've looked at the nature of wealth, especially at how new wealth is created. We learn about the creation of wealth not merely from economic theory but from experience, sometimes bitter experience. Just in case you don't have a photographic memory and are already forgetting the details, let's boil them down to the top ten ingredients in any complete recipe for creating wealth and alleviating poverty. Then, for reference, you won't have to read the whole book again. You can just flip straight to this page.

The Top Ten Ways to Alleviate Poverty; or, Creating Wealth in Ten Tough Steps

1. Establish and maintain the rule of law.

2. Focus the jurisdiction of government on maintaining the rule of law, and limit its jurisdiction over the economy and the institutions of civil society.

3. Implement a formal property system with consistent and accessible means for securing a clear title to property one owns.

4. Encourage economic freedom: Allow people to trade goods and services unencumbered by tariffs, subsidies, price controls, undue regulation, and restrictive immigration policies.

5. Encourage stable families and other important private institutions that mediate between the individual and the state.

6. Encourage belief in the truth that the universe is purposeful and makes sense.

7. Encourage the right cultural mores—orientation to the future and the belief that progress but not utopia is possible in this life; willingness to save and delay gratification; willingness to risk, to respect the rights and property of others, to be diligent, to be thrifty.

8. Instill a proper understanding of the nature of wealth and poverty—that wealth is created, that free trade is win-win, that risk is essential to enterprise, that trade-offs are unavoidable, that the success of others need not come at your expense, and that you can pursue legitimate self-interest and the common good at the same time.

9. Focus on your competitive advantage rather than protecting what used to be your competitive advantage.

10. Work hard.

These ingredients, or some subset of them, constitute the only known pathway to the large-scale creation of wealth. If we really want to alleviate poverty, we should focus on these ingredients, and not well-meaning plans that fail or do more harm than good.

Notice that, with the exception of some aspects of item nine, none of these is a material asset. You can't find economic freedom or cultural mores on a map or put them in a safe. They're intangible, immaterial, *spiritual*. They involve beliefs, social conventions, institutions, commitments, virtues, and creativity. That doesn't mean the wealthy are spiritual and the poor are unspiritual. That's the half-truth of the prosperity gospel. Here we're talking about nonmaterial things that promote wealth creation in a culture.

Christians know that the spiritual is as real as the material; that, in the ultimate scheme of things, mind precedes matter; that in the beginning was the Word. "The Spirit gives life" (John 6:63), so we should not be surprised that things spiritual are the great drivers of wealth creation in human culture.

Indeed, the more advanced an economy, the more important the intangible becomes. Recently, even the World Bank has come to appreciate this. In 2006, it released a study that highlighted the importance of so-called intangible wealth, as distinct from "natural capital" and "produced capital."

If the creation of wealth is deeply rooted in spiritual realities, we shouldn't expect unbelievers who don't think there's anything beyond the material world to readily appreciate this point. But Christians should be among the first to get it. And yet too few Christian leaders understand the one economic system that allows abundant wealth to be created—namely, capitalism. On the contrary, many prominent Christian thinkers, Protestant and Catholic alike, have made the same economic mistakes the materialists made during the previous two centuries. They've accepted the same false stereotypes as their secular counterparts—stereotypes based on anti-Christian and materialistic assumptions.

PROVIDENCE

Instead, we need to understand the mystery of the market in terms of theological categories such as providence. In Christian theology, providence refers to God's purposive preservation and guidance of all things.[1] God governs all events for a purpose,

TABLE 3. COUNTRIES WITH THE HIGHEST
PER CAPITA WEALTH, 2000

Country	Wealth per capita (US $)	Natural capital (%)	Produced capital (%)	Intangible capital (%)
Switzerland	648,241	1	15	84
Denmark	575,138	2	14	84
Sweden	513,424	2	11	87
United States	512,612	3	16	82
Germany	496,447	1	14	85
Japan	493,241	0	30	69
Austria	493,080	1	15	84
Norway	473,708	12	25	63
France	468,024	1	12	86
Belgium/ Luxembourg	451,714	1	13	86

Source: World Bank, *Where Is the Wealth of Nations? Measuring Capital for the XXI Century* (2006), 18–19, http://siteresources. worldbank.org/INTEEI/Home/20666132/WealthofNationsconfer- enceFINAL.pdf.

even if we can't discern that purpose.[2] Providence is more than simply God's knowledge of everything that will come to pass, but providence doesn't require that God determine or directly cause every event. Providence makes room for what are traditionally called "secondary causes." These include the basic properties of physical objects, such as those described by natural "laws." Providence also includes the choices of other agents God has created, such as human beings. God can act directly within nature, of course; but he is also free to work his will through these secondary causes. As the great medieval theologian Thomas Aquinas put it, God has given "even to creatures . . . the dignity of causality."[3]

God grants that dignity especially to human beings, who are free even to contradict his perfect will, to resist and even turn

TABLE 4. COUNTRIES WITH THE LOWEST PER CAPITA WEALTH, 2000

Country	Wealth per capita (US $)	Natural capital (%)	Produced capital (%)	Intangible capital (%)
Ethiopia	1,965	41	9	50
Burundi	2,859	42	7	50
Niger	3,695	53	8	39
Nepal	3,802	32	16	52
Guinea-Bissau	3,974	47	14	39
Mozambique	4,232	25	11	64
Chad	4,458	42	6	52
Madagascar	5,020	33	8	59

Source: World Bank, *Where Is the Wealth of Nations? Measuring Capital for the XXI Century* (2006), 18–19, http://siteresources.worldbank.org/INTEEI/Home/20666132/WealthofNationsconferenceFINAL.pdf.

against him. In choosing to create free creatures, God has accepted an enormous trade-off. But that doesn't mean God is powerless in the face of human sin. Rather, through his providence, God can weave together even the evil free actions of human beings into a wider, if sometimes indiscernible, greater good.

Think of the story of Joseph, who was sold into slavery by his brothers. His brothers assumed they would be rid of him forever. Most of them wished him dead. Contrary to their intentions, however, God used the event to send Joseph to Egypt, where he was eventually given a prominent position by Pharaoh. In that position, he was able to save his family, including his brothers, from famine. After the brothers meet Joseph again in Egypt, he tells them: "As for you, you meant evil against me, but God meant it for good, to bring about that many people should be kept alive, as they are today" (Gen. 50:20). Those brothers became the fathers of the twelve tribes of Israel.

Just as God could work his will through the sinful choices of Joseph's brothers, so, too, can he work his will through the free market, which involves countless trillions of individual choices, whether they be good, bad, or indifferent. No mere human can plan an economy, because no human can know all the value judgments made by the actors in an economy. But that's a limit on what we can know, not on what God can know.[4]

Rather than despising the market order, Christians should see it as God's way of providentially governing the actions of billions of free agents in a fallen world. We see the hand of God in sunrises, in pictures of Earth taken by astronauts, in mountainous landscapes and the crashing waves of the sea. We discern his fingerprints in the fine-tuning of the laws of physics and the nanotechnology inside cells. We see his handiwork in the pink shell-shaped ears of a little girl, and in a starry night sky.

Many terrible things happen in our fallen world—from tsunamis and earthquakes to malaria and death. But none of this erases the fact that, as the apostle Paul said, "ever since the creation of the world his eternal power and divine nature . . . have been understood and seen through the things he has made" (Rom. 1:20).

> "If our present order did not exist we too might hardly believe any such thing could ever be possible, and dismiss any report about it as a tale of the miraculous, about what could never come into being."
> —F. A. Hayek

Many bad things happen in a market as well, since every market exists in a fallen world. But the market order is one of the things that God has made or that emerge when his image bearers interact in a certain way. The market is, as Hayek said, "probably the most complex structure in the universe." It deserves our admiration. And yet very few Christian critics of capitalism, from liberals like William Sloan Coffin and Paul Tillich to evangelicals like Jim Wallis, Tony Campolo, and Ron Sider, have fully understood it. Fewer still have thought of it as a stunning example of God's providence over a fallen world. As we've seen, the market order is beyond the ken of man. No mere human being or committee could ever have designed it. No one anticipated or predicted it, and no one would have believed it was possible if it were not now in plain sight. It is not, as some have argued, an example of order emerging from chaos.[5] Rather, it is just what we might expect of a God who, even in a fallen world, can still work all things together for good. Seen in its proper light, the market order is as awe inspiring as a sunset or a perfect eclipse. It might not be enough to convince the *skeptic* that God exists, but the believer, at least, should see in it God's glory. At the very least, it should settle the question we started with: Can a Christian be a capitalist? The answer is surely yes.

Is the "Spontaneous Order" of the Market Evidence of a Universe Without Purpose?

Perhaps the greatest defender of capitalism in the last century was the late Austrian economist F. A. Hayek. He raised perhaps the most interesting, if somewhat technical, question for the Christian concerned about capitalism. Over the course of his career, Hayek developed a sophisticated argument against the socialist ideal of a centrally planned economy and in favor of a free-market economy structured by the rule of law.[1]

Hayek's argument focused on the mystery of the market. The mystery, as we discussed in chapter 3, is that an exquisitely complex and efficient distribution of goods and services could emerge from the (presumably) self-interested actions of individual members producing, buying, and selling in a market without fixed prices, even though none of those members sought to create such an order. In contrast, we know from communist experiments that a central planner who sets prices and determines what and how many goods and services are produced and distributed in an economy will create chaos, not order.

The great eighteenth-century thinker Adam Smith considered this "invisible hand" of the market, which transcended human limitations, as an expression of God's benevolent and providential governance of human society, since it created a more harmonious order than we would otherwise expect. If anything, we should expect chaos, and many critics of capitalism expect just that, even when they have the market order in plain sight.

Hayek was not an antireligious man, but he was an agnostic. So unlike Smith, Hayek understood the market order as what he called a "spontaneous order."[2] By definition, so his argument goes, a *spontaneous* order could never be governed by an intelligent agent. Once an activity or process reaches a certain level of complexity, it can have arisen only by spontaneous *rather than* purposeful order. Human artifacts, in comparison to the market order, are fairly simple and predictable.

Here lies the problem for socialist planning, Hayek thought. For however irreligious modern socialists may be, according to Hayek they share with primitive religions the same "naive," "superstitious," "childlike," and "animistic" belief that order requires design. This pre-Darwinian tendency to see certain kinds of order as exclusively the result of intelligent agency keeps us from recognizing the limits of economic planning on the one hand and the unique capacities of the market on the other.[3] We must recognize the limits of our reason, he insisted, and face "the implications of the astonishing fact, revealed by economics and biology, that order generated without design can far outstrip plans men consciously contrive."[4] So where does the market order come from? When push came to shove, Hayek concluded that it is an example of order emerging from chaos.

Hayek's argument, then, seems to indict not just socialist planning, but the idea of a purposeful universe in general. Both, according to Hayek, are fruit from the same poisoned tree. Hayek saw a market with its coordinated prices as the "result of human action but not of human design."[5] In other words, the market order, like language and human custom, emerges from the interaction of human beings with each other and the environment, but none is reducible either to human reason or to design. Languages, for instance, aren't created by committee. Markets, Hayek argued, are like that. As they become more complex, they become less amenable to conscious direction by one or several intelligent agents.

If he's right, then a key argument I've used in this book is also an argument against the cosmic purpose Christians believe in. But does his argument work? As an argument against central planning of an economy, it does. As an account of the origin of the market order, it does not. Most of the mischief is in his use of the fuzzy word *order*. He uses the word *order*, in fact, to cover fundamentally different realities.[6] At one end of the spectrum is the order common in physics and chemistry, much of which can be accurately described by certain mathematical laws. We appeal to the force of gravity, for instance, to predict the orbit of the planets around the sun, the movement of a projectile above a field, or how long it will take for a falling metal ball to get from the top of the Leaning Tower of Pisa to the ground. We are talking here about natural regularities, which occur in regular and predictable ways.

Other kinds of order are quite different, however, and are almost the opposite of the regularities that we find in physics. The literal opposite of order is randomness or chaos. Mere randomness is a type of complexity, but it is not an orderly complexity. By definition, randomness is irregular and unpredictable.[7]

A meaningful message is also irregular and unpredictable, but it's not random.

The information in a message, unlike the order we find in physics, is irregular, unpredictable, and complex; but unlike something that exhibits randomness, a message is specified, since it conforms to some meaningful pattern. If the message is in English, the words acquire specific meanings as they conform to the rules of English grammar and syntax, providing a higher-order meaning not present in the letters or individual words themselves. We find this type of semantic "order" in everything from human language and computer software to certain "coding regions" of DNA. We normally call this type of order *information*.

Think of the difference between Scrabble letters tossed thoughtlessly on the floor and the same Scrabble letters sequenced carefully to convey a meaningful message, such as "HAPPY TWENTY FIFTH ANNIVERSARY." Both the random pile and the intentionally ordered sequence are complex and irregular (or "aperiodic"). But the former specifies nothing in particular, while the latter conforms to the rules of grammar and syntax and contains semantic information.

Now compare chaos and information on the one hand with mere physical order on the other. Physical order is like a repetitive series of Scrabble letters, such as ABABABABABABABA. That is certainly orderly, but it is hardly information-rich.

In contrast, the DNA molecule can contain information for building proteins in the sequences of nucleotide pairs only because no physical law determines the sequence of those pairs. If the pairs had a strong chemical attraction to each other (along the information-bearing axis of the molecule), they would not be a good medium for carrying information. Similarly, we can transmit information by modulating—that is, varying—electromagnetic waves only because those waves otherwise conform to regular patterns—in this case, specific wavelengths. If such wavelengths varied randomly or couldn't be modulated, they would be bad carriers of information.

In the same way, we can use ink and paper to write a letter only because ink can be applied to paper in countless different ways, and they don't react randomly. Notice in these examples that we have passed from the simple order of physics to the complex orders of molecular biology, human language, and artifacts. The order of information is neither simple and repetitive, like the order of physics, nor random. It is a third type of reality.

Now, the "order" of the market is much more like the information we find in language, molecular biology, and computer technology than the simple recurring order we find in physics. So analogies drawn between systems in the physical sciences and the market order are bound to be weak. And here lies a problem with Hayek's analysis of spontaneous order in the market. For it is in simple physical systems that we have the clearest examples of spontaneous order, or what is often called self-organization. Highly symmetrical patterns emerge in quartzes, for instance, simply as the result of the chemical properties of the elements that make them up. And we have all seen the highly geometric patterns of snowflakes, which result from the chemical properties of water when it is frozen. Though these orderly patterns might suggest some wider purpose in the universe (for example, at the level of the fine-tuning of physical constants that describe how matter interacts with other matter), we understand that snowflakes can emerge without any direct input from an intelligent agent. But this type of order is highly symmetrical, repetitive, and simple compared with, say, the order of the genetic information that cells use to assemble proteins. Even if we do not need to invoke design directly to explain a snowflake or quartz, it doesn't follow, as Hayek seems to assume, that we could not invoke design for the fundamentally different type of order found everywhere from literature to biology to economics.

When we move to the biological realm, of course, things get more complicated. Hayek sees many parallels between the biological and economic realms. He argues that modern biology borrowed some of its most significant ideas, especially evolution, from economists.[8] And although he goes well beyond Darwinian natural selection in describing the evolution of human traditions and of the "extended order of the market," he nevertheless argues that certain traditional rules have survived in successful cultures because they confer a survival advantage on that culture.[9]

CAN ORDER EMERGE FROM CHAOS?

Hayek had many harsh things to say about the Greek philosopher Aristotle. In criticizing Aristotle's "utter incomprehension of the market order in which he lived," Hayek argued that Aristotle "lacked any perception of two crucial aspects of the formation of *any complex structure* [emphasis added], namely, evolution and the self-formation of order. As Ernst Mayr . . . puts it, 'The idea that the universe could have developed from an original chaos, or that higher organisms could have evolved from lower ones, was totally alien to Aristotle's thought.'"[10]

Notice Mayr's claim, which Hayek quotes with approval, *that the universe could have developed from an original chaos.* Hayek seems to take this idea as a rough equivalent to his "self-formation of order"—that is, to *spontaneous order.* Hayek's argument is something like this: *We already accept that order can emerge from chaos in cosmology and biology; therefore,*

we have no good reason to deny the same phenomenon in eco-
nomics. Therefore, we have no reason to appeal to purpose in
explaining the market order.

But has modern science *really* discovered that order emerges
from chaos? In fact, modern cosmologists have discovered just
the opposite. Cosmologists of various metaphysical stripes
now agree that the initial conditions present at the beginning
of cosmic history must have been in a state of extraordinary
order—low entropy—with matter conforming to very precise
physical laws. It's so orderly, in fact, that it raises the question
of who or what caused it all. Physicists and cosmologists refer
to this as the "fine-tuning problem," because the universe, by all
appearances, was fine-tuned for the possibility of life. So in cosmology there is no evidence of order emerging from chaos, but
rather order from order.[11]

The origin of life is not well understood, but again, given both
life's rarity and its exquisitely ordered complexity, we have no
reason to assume that it emerged from chaos, and every reason
to assume that, at a minimum, it requires very precise initial
conditions. In biology—that is, after the origin of life—we enter
a higher order of complexity than in physics and chemistry. We
are now dealing with organisms, which resist simple mathematical explanations. As chemist Arthur Robinson has said: "Using
physics and chemistry to model biology is like using Lego blocks
to model the World Trade Center."[12]

And however generous we are with the Darwinian mechanism, it doesn't illustrate order emerging from chaos. On the
contrary, Darwin proposed natural selection and random variation as a *mechanism* for producing adaptive complexity without
design. He didn't suggest random variation alone. In fact, he
treated mere chance or chaos as too improbable to entertain, and
most Darwinists have followed him in that assessment.[13]

From biology we move to the human sciences. Here the effects of intelligent agents appear everywhere. So it's no surprise
that it's harder to use math to model human behavior than it is
to use it to model, say, the movement of a ball rolling down a

hill. By the time we reach economics, we are dealing not only with human agents, but with the complexity of the market exchanges of millions or billions of intelligent agents. As we go from physics at one end to economics at the other, we are moving up a "nested hierarchy" of complexity, in which higher orders constrain but cannot be reduced to lower orders.[14] Nowhere along the ladder do we find evidence of order emerging from chaos.

In any case, we have no reason to assume, let alone require, that the type of order we find in the economic realm must be reducible to the order we find in physics or biology. It could be an order unto itself.

Hayek should have seen this. "So far as we know," he said, "the extended order is probably the most complex structure in the universe."[15] Breaking with more consistent materialists, he recognized the immateriality of economics in his subjective theory of value: "An increase in value . . . is indeed different from increases in quantity observable by our senses."[16] And he discerned that human agency couldn't be reduced to physics: "In a certain sense the activity that economics sets out to explain is not *about* physical phenomena but about people."[17]

But in the end, Hayek observed the well-known boundaries of academic respectability, which call for strict materialism.[18] Therefore, he tried to justify his concept of spontaneous order by appealing to processes safely within the natural sciences,[19] which, at other moments, he knew were inadequate. He could not appeal to God, an eternal universe, or some purpose beyond the physical universe, so he opted for the remaining option: the market order must be an example of order emerging from chaos. But we have no reason to think that principle is true,[20] whether in physics, chemistry, biology, or economics.

Besides, the market order doesn't just appear from nowhere. It happens only under the right conditions, including the rule of law and private property. And individual agents transmit and carry each piece of information from beginning to end. Human choices shape the lower-level trajectories. It's the overall order

that, from the perspective of the individual participants, is surprising,[21] even gratuitous. But this market "spontaneity" is no
argument against a wider cosmic purpose.[22]

In fact, it makes a lot more sense in a providential—a
purposeful—universe.

The ideas for this book began many years ago, and come from so many sources that any list of acknowledgments will be incomplete. I benefited from the insights of many in the process of actually writing the book. I would especially like to thank Jonathan Witt for his careful reading and editorial suggestions of early drafts, and Kaylin Wainright and Alex Binz for their research assistance. For their help and insights, I would like to thank Rev. Robert Sirico, Kris Mauren, Michael Miller, Sam Gregg, Bob Cihak, George Gilder, Gina and Glenn Overweg, and Ginny Richards. Finally, I am grateful to my agent, Giles Anderson, and to my editors, Roger Freet and Lisa Zuniga, who helped me improve the manuscript substantially, so that I only rarely use boring adjectives like "substantially."

INTRODUCTION: CAN A CHRISTIAN BE A CAPITALIST?
1. Rich Karlgaard, "Godly Work," *Forbes,* April 23, 2007.

CHAPTER 1: CAN'T WE BUILD A JUST SOCIETY?
1. At the time, I thought the deal was that five readings of the book would equal an "A" for the course. Later, as I thought back on the course, it occurred to me that I may have misunderstood the assignment. Professor Kim may have said that if we read the text five times, we would be bound to get an "A" on the exams.
2. Richard Pipes, *Communism: A History* (New York: Modern Library Chronicles, 2003), 17. In England, real income per capita doubled from 1760 to 1860. See Nicholas F. R. Crafts, *British Economic Growth During the Industrial Revolution* (Oxford: Clarendon Press, 1985). See also the discussion in Thomas Woods, *The Church and the Market: A Catholic Defense of the Free Economy* (Lanham, MD: Lexington Books, 2005), 169–74.
3. Pipes, *Communism,* 38–39.
4. Pipes, *Communism,* 45.
5. Pipes, *Communism,* 67.
6. For the excruciating details, see Jean-Louis Margolin, "China: A Long March into Night," in *The Black Book of Communism: Crimes, Terror, Repression,* by Stéphane Courtois et al. (Cambridge, MA: Harvard Univ. Press, 1999), 463–546. Modern China is growing despite Mao's vision because of free-market reforms introduced by Deng Xiaoping in 1979 and continuing to the present.
7. Edward S. Herman and Noam Chomsky, "Distortions at Fourth Hand," *Nation,* June 25, 1977, 789–93. See also the story "Remembering the Killing Fields," *CBS News,* April 2000, http://www.cbsnews.com/stories/2000/04/15/world/main184477.shtml.

8. Jean-Louis Margolin, "Cambodia: The Country of Disconcerting Crimes," in Courtois et al., *Black Book of Communism*, 588–95; Pipes, *Communism*, 133–35.
9. Martin Malia, "Foreword: The Uses of Atrocity," Courtois et al., *Black Book of Communism*, x.
10. Ronald J. Sider, *Rich Christians in an Age of Hunger: Moving from Affluence to Generosity*, 20th anniversary revision (Nashville: Word Publishing, 1997), 78.
11. For a detailed history of communism vis-à-vis socialism, see Joshua Muravchik, *Heaven on Earth: The Rise and Fall of Socialism* (San Francisco: Encounter Books, 2003).
12. Tom Bethell, *The Noblest Triumph: Property and Prosperity Through the Ages* (New York: Palgrave Macmillan, 1999), 37–45.
13. See Alvaro Vargas Llosa, *The Che Guevara Myth and the Future of Liberty* (Oakland, CA: Independent Institute, 2006).
14. Quoted by Martin Malia in Courtois et al., *Black Book of Communism*, xx.
15. As Paul Johnson put it in *Modern Times* (New York: Harper-Collins, 1991), 544. "Experimenting with Half Mankind" is the title of Johnson's chap. 16, which treats Mao's Communist Revolution in China.
16. See Johnson's discussion, *Modern Times*, 546ff.
17. *Red Flag* (Beijing), June 1, 1958.
18. Rev. 11:15, 19:11–16. Paul also speaks of the kingdom in future terms: 1 Cor. 6:9, 10, 15:50; Gal. 5:21; Eph. 5:5; 1 Thes. 2:12; 2 Thes. 1:5; Col. 4:11; 2 Tim. 4:1, 18. See J. Ramsey Michaels, "The Kingdom of God and the Historical Jesus," in *The Kingdom of God in 20th-Century Interpretation,* ed. Wendell Willis (Peabody, MA: Hendrickson Publishers, 1987), 112.
19. Nirvana is sort of like the Buddhist concept of salvation. Strictly speaking, it refers to a state in which one is freed from or extinguished from the eternal cycle of birth, death, and rebirth, a cycle that gives rise to suffering. Many ordinary Buddhists view Nirvana as a sort of heavenly place where souls go after death.

CHAPTER 2: WHAT WOULD JESUS DO?

1. For a discussion of the virtues of prudence, see Joseph Pieper, *The Four Cardinal Virtues* (South Bend, IN: Univ. of Notre Dame Press, 1990), 6–9.

2. Henry Hazlitt, *Economics in One Lesson* (New York: Three Rivers Press, 1979), 17.
3. A transcript and an MP3 audio titled "Jim Wallis: God Hates Inequality" are available. See the *Huffington Post,* February 1, 2007, or http://www.huffingtonpost.com/jim-wallis/god-hates-inequality-b–40170.hml.
4. Hazlitt, *Economics in One Lesson,* 136.
5. See James Sherk, "Union Members, Not Minimum-Wage Earners, Benefit When the Minimum Wage Rises," *Heritage Foundation WebMemo 1350* (February 2, 2007), http://www.heritage.org/Research/Economy/wm1350.cfm.
6. See Jordan Ballor, "Strange Brew: Churches Push for Fair Trade Coffee," *Acton Commentary* (February 3, 2004): http://www.acton.org/commentary/commentary_178.php. See links to the various religious "fair trade" organizations in Ballor's article. There is a lot of economic confusion in these organizations' statements. In this section, however, I'm presenting the best case for the fair-trade concept, even though many "fair trade" organizations trade in simple economic confusions.
7. Companies like Starbucks have helped transform coffee from a commodity (where one bean is worth about the same as every other bean) into a highly diversified industry where coffee is a luxury item. As a result, Starbucks and other espresso companies often pay near or above the "fair trade" price for high-end beans. See Kerry Howley's excellent article "Absolution in Your Cup," *Reason,* March 2006, http://reason.com/news/show/33257.html.
8. This would have the same problems as does every attempt to fix prices above market price: waste and overproduction. See, for instance, the comments of Global Exchange at http://www.globalexchange.org/campaigns/fairtrade/coffee/index.html. Although attempts to get the government to establish "fair trade" prices are not likely to succeed anytime soon, many fair-trade organizations do badger coffee companies to carry "fair trade" certified coffee. It's not clear how many would do so if only market factors were involved. What is clear is that "fair trade" coffee and other products make up only 1 to 3 percent of the total market. For a more detailed analysis of the economic consequences of "fair trade" policies, see Brink Lindsey, *Grounds for Complaint? "Fair Trade" and*

the Coffee Crisis (London: Adam Smith Institute, 2004), http://www.adamsmith.org/pdf/groundsforcomplaint.pdf.

9. I think fair trade is best understood as a mixed act—part ordinary trade, part charity. So I disagree with Alex Nicholls and Charlotte Opal in *Fair Trade: Market-Driven Ethical Consumption* (London: Sage Publications, 2005), 13. See this book for a defense of fair trade and for a discussion of various fair-trade organizations.

10. You can get all sorts of boring statistics on the world coffee market from the International Coffee Organization at http://www.ico.org/.

11. One of many is TechnoServe. See it online at http://www.technoserve.org/. Of course, Christian missionaries seeking to spread the gospel may be the best medicine of all over the long term, since property laws can grow only so well without a government and people who respect the property of others.

12. Paul Collier, *The Bottom Billion: Why Poor Countries Are Failing and What Can Be Done About It* (Oxford: Oxford Univ. Press, 2007), 104.

13. See http://action.one.org/issues/.

14. David Neff, "UN Leader Woos Evangelicals," *Christianity Today liveblog,* October 12, 2007, http://blog.christianitytoday.com/ctliveblog/archives/2007/10/un_leader_woos.html.

15. Quoted in Gethin Chamberlain, "Nice Concert. But Can It Really Save Millions from Dying?" *Scotsman,* July 4, 2005.

16. William Easterly, "Foreign Aid Face Off," *Los Angeles Times,* April 20, 2006.

17. Raghuram G. Rajan and Arvind Subramanian, "Aid and Growth: What Does the Cross-Country Evidence Really Show?" *NBER Working Paper 11513* (August 2005). See abstract at http://papers.nber.org/papers/w11513.

18. Foreign aid for development is different from postwar aid programs, like the Marshall Plan, which are given to already developed countries (like Germany after World War II) that have most of the cultural and legal institutions in place for channeling the aid appropriately. It is also different from emergency help after natural disasters, such as the help that the U.S. and Australian militaries provided to Indonesia immediately after the tsunami of 2004.

19. Easterly, "Foreign Aid Face Off."

20. For detailed documentation, see Amity Shlaes, *The Forgotten Man: A New History of the Great Depression* (New York: Harper-Collins, 2007).

21. Carmen DeNavas-Walt, Bernadette D. Proctor, and Cheryl Hill Lee, *Income, Poverty, and Health Insurance Coverage in the United States: 2005*, U.S. Census Bureau, Current Population Reports, P60–231 (Washington, DC: Government Printing Office, 2006), 13, http://www.census.gov/prod/2006pubs/p60–231.pdf.

22. These figures are drawn from Charles Murray, *Losing Ground: American Social Policy, 1950–1980,* 10th anniversary ed. (New York: Basic Books, 1994); D. Eric Schansberg, *Poor Policy: How Government Harms the Poor* (New York: Westview, 1996); and Michael Tanner, *The Poverty of Welfare: Helping Others in Civil Society* (Washington, DC: Cato Institute, 2003).

23. George A. Akerlof and Janet L. Yellen, "An Analysis of Out-of-Wedlock Birth in the United States" (Washington, DC: Brookings Institution, 1996), http://www.brook.edu/comm/policybriefs/pb05.htm.

24. Pius XI, *Quadragesimo Anno* (1931).

25. Clint Bolick, "A Cheer for Judicial Activism," *Wall Street Journal,* April 3, 2007.

26. John Stossel, *Give Me a Break* (New York: HarperCollins, 2004), 131–32.

27. Peter Drucker, *Post-Capitalist Society* (New York: HarperCollins, 1993), 123.

28. George Gilder, *Wealth and Poverty* (San Francisco: ICS Press, 1993), 30.

29. E. J. Dionne, "The Overlooked Schism: America's Religious Communities and the Battle over Government," *Sojourners,* April 2007, http://www.sojo.net/index.cfm?action=magazine.article&issue=soj0704&article=070410.

30. Jim Wallis, "All Hands on Deck: Scripture Suggests a Clear Role for Government in Ensuring the Common Good," *Sojourners,* April 2007, http://www.sojo.net/index.cfm?action=magazine.article&issue=soj0704&article=070451.

CHAPTER 3: DOESN'T CAPITALISM FOSTER UNFAIR COMPETITION?

1. Richard Hofstadter, *Social Darwinism in American Thought,* rev. ed. (Boston: Beacon Press, 1955), 5, quoted in John West, *Darwin*

Day in America: How Our Politics and Culture Have Been De-humanized in the Name of Science (Wilmington, DE: ISI Books, 2007), 107. Hofstadter popularized the idea that nineteenth-century American capitalists defended capitalism by appealing to social Darwinism. West shows that Hofstadter's thesis is largely untrue, and that such arguments were mostly restricted to Darwinist intellectuals like Sumner and Herbert Spencer rather than to actual businessmen. Many of the myths about the nineteenth-century "robber barons" are just that—myths. For illuminating discussion, see Burton W. Folsom Jr., *The Myth of the Robber Barons,* 3rd ed. (Herndon, VA: Young America's Foundation, 1996).

2. Jack London, "The Class Struggle," in *War of the Classes* (Oakland: Star Rover House, 1982), 18, quoted in West, *Darwin Day in America,* 113.

3. Walter Mondale, *New York Times,* February 22, 1983.

4. Pius XI, *Quadragesimo Anno,* 88.

5. Smith leads with this idea of valuing things in terms of labor or cost of production in *An Inquiry into the Nature and Causes of the Wealth of Nations,* ed. Edwin Cannan (New York: Modern Library, 1994), lix. Nevertheless, the idea was more central to Marx's thought than to Smith's.

6. Danny Duncan Collum, "We Buy, Therefore We Are: Everyday Low Prices Are Part of Our American Birthright. Right?" *Sojourners,* March 2007, http://www.sojo.net/index.cfm?action=magazine. article&issue=soj0703&article=070340.

7. Marx tried to get around these problems by defining economic value in terms of "socially necessary labor." But that just moves the problem back a step. What is socially necessary? If we say labor is socially necessary if it produces some good or service that someone values, then the labor theory collapses into the "subjective theory" of economic value. So Marx's qualification of his labor theory of value, if followed consistently, guts the theory itself.

8. Jim Wallis, "God Hates Inequality," *Huffington Post,* February 1, 2007, http://www.huffingtonpost.com/jim-wallis/god-hates-inequality_b_40170.html.

9. The subjective theory of economic value is often associated with the Austrian economists such as Carl Menger, Ludwig von Mises, and F. A. Hayek. But medieval Christian scholars anticipated most of the Austrians' insights in their commentaries on Aristotle's

Ethics. See Alejandro A. Chafuen, *Faith and Liberty: The Economic Thought of the Late Scholastics* (Lanham, MD: Lexington Books, 2003); and Odd Langholm, *Price and Value in the Aristotelian Tradition* (Oslo: Universitetsforlaget, 1979). These Scholastic insights were most fully developed by the Spanish School of Salamanca, 1544–1605. See Marjorie Grice-Hutchison, *The School of Salamanca* (Oxford: Clarendon Press, 1952).

10. Carl Menger, *Principles of Economics* (New York: New York Univ. Press, 1981).

11. Quoted in Andrea Gabor, "Deirdre McCloskey's Market Path to Virtue," *Strategy + Business* 43 (Autumn 2006): 8.

12. In the Christian tradition, Thomas Aquinas provided the classical defense of private property in *Summa Theologica* II-II, Q. 66, art. 2. For a sophisticated treatment of the balance of private property with the doctrine of the "universal destination of goods," see Manfred Spieker, "The Universal Destination of Goods: The Ethics of Property in the Theory of a Christian Society," *Journal of Markets and Morality* 8, no. 2 (Fall 2005): 333–54.

13. Gerhard von Rad, *Genesis* (Philadelphia: Westminster Press, 1972), 249. Von Rad attributes this passage to the "Priestly" source, and assumes it would have been included only if it served a specific theological purpose. He thinks its purpose is to show that the great patriarch Abraham, though he had forsaken everything, came to own a small part of the Promised Land before his death. My argument, of course, is not that the author of Genesis included this passage to argue that the ancient Near East had a highly developed concept of land titling. Quite the contrary. That fact emerges implicitly, much like the other examples of private property in the Bible. The concept is everywhere assumed rather than defended explicitly. Biblical scholars generally agree that this passage describes the ancient Near Eastern equivalent of a legal titling of a plot of land. For example, see E. A. Speiser, *Genesis* (Garden City, NY: Doubleday, 1986), 171.

14. Hernando de Soto, *The Mystery of Capital: Why Capitalism Triumphs in the West and Fails Everywhere Else* (New York: Basic Books, 2000).

15. De Soto, *Mystery of Capital,* 35, 36.

16. Hernando de Soto, "Why Capitalism Works in the West but Fails Elsewhere," *International Herald Tribune,* January 5, 2001.

17. De Soto, *Mystery of Capital,* 19.

18. De Soto, *Mystery of Capital*, 21, 26–27.

19. De Soto, *Mystery of Capital*, 40. De Soto estimates that as much as 70 percent of the credit received for new businesses "comes from using formal titles as collateral for mortgages" (84).

20. For more on the importance of private property, see Tom Bethell, *The Noblest Triumph* (New York: St. Martin's Press, 1998).

21. Smith, *Wealth of Nations*, 485.

22. The most mature form of Hayek's argument is in *The Fatal Conceit: The Errors of Socialism* (Chicago: Univ. of Chicago Press, 1989).

23. Hayek, *Fatal Conceit*. Imagine how hard it would be for a benevolent planner in Washington, D.C., to guess how much I would pay for two balcony tickets to see the Bare Naked Ladies concert (the Canadian band) on July 12, 2010, in Van Andel Arena, Grand Rapids, Michigan. Better yet, how much would I pay for two more tickets, after I'd already bought two? Even I couldn't answer that question right now, so no one else could, either. Now, that's just one teensy-weensy bit of information. No human being, or human committee, can ever have access to the total information needed to plan an economy. See also F. A. Hayek, *The Use of Knowledge in Society* (Arlington, VA: Institute for Humane Studies, 1977). For a discussion of how Hayek's argument developed over time, see Bruce Caldwell, "Hayek and Socialism," *Journal of Economic Literature* 35, no. 4 (December 1997): 1856–90.

24. Economist Deidre McCloskey makes this point nicely in "What Would Jesus Spend? Why Being a Good Christian Won't Hurt the Economy," *In Character,* Fall 2004, http://www.incharacter.org/article.php?article=8.

25. Leonard E. Read, *I, Pencil* (Irving-on-Hudson, NY: Foundation for Economic Education, 2006).

26. Quoted in Julian Simon, *The Ultimate Resource 2* (Princeton, NJ: Princeton Univ. Press, 1996), 610.

27. Tim Manners, "iPod Math," *Reveries,* July 5, 2007, http://reverie.com/?p=1170.

28. A recent example is Kathryn Tanner, *Economy of Grace* (Minneapolis: Augsburg Press, 2005). Tanner's cartoonishly simple contrast between an "economy of competition" and an "economy of grace" exhibits an all too common deficiency among theologians who comment on economics: they know virtually nothing about economics beyond the ambient caricatures picked up in the

academy. She also ignores the destructive consequences that would surely follow from the policies she advocates.

CHAPTER 4: IF I BECOME RICH, WON'T SOMEONE ELSE BECOME POOR?

1. Mark Steyn, *America Alone* (Washington, DC: Regnery, 2006), xi.

2. "At First Meeting, Christian Churches Together Tackles Poverty," *Catholic News Service,* February 15, 2007.

3. Wallis, "God Hates Inequality."

4. Bishop Thomas Gumbleton, in the documentary film *The Call of the Entrepreneur* (Acton Media, 2007).

5. John Paul II, "Homily in the Jose Marti Square of Havana, January 25, 1998," www.vatican.va/holy_father/john_paul_ii/travels/documents/hf_jp-ii_hom_25011998_lahavana_en.html; quoted in Johan Norberg, *In Defense of Global Capitalism* (Washington, DC: Cato Institute, 2003), 20.

6. Gustavo Gutierrez, *A Theology of Liberation* (Maryknoll, NY: Orbis Books, 1994). The book was originally published in English in 1973. Gutierrez has retracted some of his more extreme and clearly false claims about "dependency theory" in later editions of the book. This may have been in response to the devastating critique of Michael Novak in *Will It Liberate? Questions About Liberation Theology* (Lanham, MD: Madison Books, 2000). But even in the most recent edition of his book, Gutierrez retains the zero-sum, "gap" reasoning that presupposes dependency theory. See, for instance, Gutierrez, *Theology of Liberation,* 53.

7. Sider, *Rich Christians,* 144.

8. Robert E. Rector and Kirk A. Johnson, *Understanding Poverty in America* (Washington, DC: Heritage Foundation, 2004), http://www.heritage.org/Research/Welfare/bg1713.cfm.

9. At http://www.millenniumcampaign.org/sitepp.asp?c=grKVL2NLE&b=185518. References for these statistics are available at this Web site.

10. Luisa Kroll and Allison Fass, "The World's Billionaires," *Forbes,* March 8, 2007, http://www.forbes.com/2007/03/07/billionaires-worlds-richest_07billionaires_cz_lk_af_0308billie_land.html.

11. Actually, there are some troubling aspects to Helu's business practices, since he seems to have gamed the legal system in Mexico. But just substitute him for another billionaire whose gain is less ill begotten, and the point is the same.

12. Unlike the Millennium Campaign, which rightly focuses most of its attention on absolute poverty, other departments of the United Nations seem fixated on income inequality and gap and zero-sum thinking when it comes to development, rather than on poverty itself. See, for instance, the U.N. Department of Economic and Social Affairs report *The Inequality Predicament: Report of the World Social Situation 2005*, p. 1: "Focusing exclusively on economic growth and income generation as a development strategy is ineffective, as it leads to the accumulation of wealth by a few and deepens the poverty of many." Available at: http://www.ilo.org/public/english/region/ampro/cinterfor/news/inf_05.pdf.

13. U.S. Census Bureau, "Historical Income Tables—Families," http://www.census.gov/hhes/www/income/histinc/f01ar.html.

14. P. T. Bauer shows that the countries least touched by markets are the poorest. P. T. Bauer, *Equality, The Third World, and Economic Delusion* (Cambridge, MA: Harvard Univ. Press, 1981), 67–68. See the discussion in Michael Novak, *The Spirit of Democratic Capitalism* (Lanham, MD: Madison Books, 1991), 109.

15. You can see this illustrated by going to www.gapminder.org, clicking on "Human Development Trends," and then clicking on presentation 3, "Poverty."

16. See, for instance, the recent study by Indur M. Goklany, *The Improving State of the World: Why We're Living Longer, Healthier, More Comfortable Lives on a Cleaner Planet* (Washington, DC: Cato Institute, 2007).

17. See Mary Anastasia O'Grady, "The Poor Get Richer," *Wall Street Journal,* January 16, 2007. View and download the *Index of Economic Freedom,* ed. Mary Anastasia O'Grady, Tim Kane, and Kim R. Holmes, at http://www.heritage.org/index/.

18. Hernando de Soto, "The Mystery of Capital," *Finance and Development* 38, no. 1 (March 2001).

19. De Soto, *Mystery of Capital,* 204.

20. De Soto, *Mystery of Capital,* 157.

21. De Soto, *Mystery of Capital,* 177.

22. De Soto, *Mystery of Capital,* 7.

23. Rafael Di Tella, Sebastian Galiani, and Ernesto Schargrodsky, "Property Rights and Beliefs: Evidence of the Allocation of Land Titles to Squatters," *Quarterly Journal of Economics,* February 2007. See also the review article by Julia Hanna, "How Property

Ownership Changes Your World View," *Harvard Business School: Working Knowledge,* May 28, 2007.

24. See the discussion in Victor P. Hamilton, *The Book of Genesis, The New International Commentary on the Old Testament* (Grand Rapids, MI: Eerdmans, 1990), 134–35. See also von Rad, *Genesis,* 57–58.

25. Generic "man" in Hebrew is *adam,* which is related to the word "ground," *adamah.*

26. There are some complicated issues in biblical interpretation here. Some scholars insist that Cain and Abel were practicing full-blown agricultural domestication. Others see Cain and Abel as simply representing these two later groups—hunters and gatherers. Still others suggest that Cain and Abel were practicing a primitive form of semidomestication, practices that blur the line between hunting/ gathering and thorough-going domestication, practices for which there were no precise Hebrew words when Genesis was written. Regardless of which perspective one prefers, it's clear that Cain and Abel function as the representatives of these two main types of agricultural creators.

27. De Soto, *Mystery of Capital,* 40.

28. Michael Novak, *The Spirit of Democratic Capitalism,* rev. ed. (Lanham, MD: Madison Books, 2000).

29. Richard L. Stroup, *Eco-nomics: What Everyone Should Know About Economics and the Environment* (Washington, DC: Cato Institute Press), 10.

30. See Ray Kurzweil, *The Age of Spiritual Machines* (New York: Penguin, 2000). For discussion and criticism of some of Kurzweil's ideas, see Jay Richards and George Gilder, eds., *Are We Spiritual Machines?* (Seattle: Discovery Institute Press, 2002).

31. Commentators often complain about multinational and transnational corporations getting rich by using low-wage labor in developing countries. Sometimes there is injustice involved, such as when land is confiscated without just compensation. But the fact that, say, an American company pays a factory worker less in Guatemala than it would pay a worker in Ohio isn't unjust. If the company were forced to pay such workers the same amount, it would have no incentive to put a factory in Guatemala in the first place, since the company incurs far greater risk in putting a factory in Guatemala. Moreover, for a variety of reasons, the labor is

often less productive (and so less valuable) in third-world countries than in the more developed countries. Finally, most American and European-based companies pay their workers in developed countries far more than comparable locally owned companies. But even if such an arrangement were unjust, it still wouldn't follow that the Guatemalan workers are getting *poorer* as a result of the American company getting richer.

32. Pope Eugene IV condemned slavery in 1435 in *Sicut Dudum* (available at http://www.papalencyclicals.net/Eugene04/eugene04sicut .htm), and later popes followed him. But the worldwide abolition of slavery did not become a practical reality until the nineteenth century.

33. Of course, Malachi is clearly appealing to fellow Jews in this passage and not to humanity in general. And he's not addressing equality per se. But all human beings are all created by the same God, and all are made in his image.

34. Sider, *Rich Christians,* 144.

35. This is the most current data available at the time of this writing, updated in September 2006; but it is consistent with other years. See Gerald Prante, "Summary of Latest Income Data," http:// www.taxfoundation.org/taxdata/show/250.html. Of course, even a flat income tax fixed at 1 percent for all taxpayers will lead to more tax revenue coming from higher income brackets.

36. Robert A. Guth, "How Bill Gates Got Ready for Harvard," *Wall Street Journal,* June 8, 2007, 1.

CHAPTER 5: ISN'T CAPITALISM BASED ON GREED?

1. Michael Sieply, "Film's Wall Street Predator to Make a Comeback," *New York Times,* May 5, 2007.

2. Tony Campolo, *Letters to a Young Evangelical* (New York: Basic Books, 2006).

3. Bernard Mandeville, *The Fable of the Bees*, with commentary by F. B. Kaye (Oxford: Oxford Univ. Press, 1924; repr., Indianapolis: Liberty Fund, 1988), 1:24–26.

4. Mandeville, *Fable of the Bees,* 369.

5. An article such as Rose Marie Berger, "What the Heck Is 'Social Justice'?" *Sojourners,* February 2007, http://www.sojo.net/index .cfm?action=magazine.article&issue=soj0702&article=070265.

6. Jena McGregor, "Sweet Revenge," *Business Week,* January 22, 2007, 66.

7. Smith, *Wealth of Nations,* xliii.
8. Smith, *Wealth of Nations,* 148.
9. Smith, *Wealth of Nations,* 15.
10. Smith, *Wealth of Nations,* 485.
11. With Nathaniel Branden (New York: Signet, 1964).
12. Ayn Rand, *Capitalism: The Unknown Ideal* (New York: Signet, 1967), 195.
13. Ayn Rand, *Atlas Shrugged* (New York: Random House, 1957), app.
14. Daniel J. Flynn, *Intellectual Morons: How Ideology Makes Smart People Fall for Stupid Ideas* (New York: Crown Forum, 2004), 200–201.
15. See the excellent article on this point by Robert A. Black, "What Did Adam Smith Say About Self-Love?" *Journal of Markets and Morality* 9, no. 1 (Spring 2006): 7–34.
16. See also Jerry Z. Muller, *Adam Smith in His Time and Ours* (New York: Free Press, 1993), 71.
17. The "butcher, brewer, baker" quote is notoriously misinterpreted when pulled out of context. The larger quote makes Smith's point more clearly: "But man has almost constant occasion for the help of his brethren, and it is in vain for him to expect it from their benevolence only. He will be more likely to prevail if he can interest their self-love in his favour, and shew them that it is for their own advantage to do for him what he requires of them. Whoever offers to another a bargain of any kind, proposes to do this. Give me that which I want, and you shall have this which you want, is the meaning of every such offer; and it is in this manner that we obtain from one another the far greater part of those good offices which we stand in need of. It is not from the benevolence of the butcher, the brewer, or the baker that we expect our dinner, but from their regard to their own interest. We address ourselves, not to their humanity, but to their self-love, and never talk to them of our own necessities, but of their advantages" (*Wealth of Nations,* 15).
18. So Smith, in his *Theory of Moral Sentiments,* says: "It is the great fallacy of Dr. Mandeville's book to represent every passion as wholly vicious which is so in any degree and in any direction." Quoted in F. B. Kaye's commentary to Mandeville, *Fable of the Bees,* 2:414. Duncan Foley seems unaware of this, and so projects Mandeville's views onto Smith in his recent book *Adam's Fallacy: A Guide to Economic Theology* (Cambridge, MA: Harvard Univ. Press, 2006): "The moral fallacy of Smith's position is that

it urges us to accept direct and concrete evil in order that indirect and abstract good may come of it" (3). Throughout the book Foley confuses Smith's idea of self-interest with the vice of selfishness.

19. "How selfish soever man may be supposed, there are evidently some principles in his nature, which interest him in the fortune of others, and render their happiness necessary to him, though he derives nothing from it except the pleasure of seeing it." Smith, *Theory of Moral Sentiments*, 9.

20. Adam Smith, *The Theory of Moral Sentiments*, ed. D. D. Raphael and A. L. Macfie (Oxford: Oxford Univ. Press, 1976; reprint Indianapolis: Liberty Fund, 1981); quoted in P. J. O'Rourke, *On the Wealth of Nations* (New York: Atlantic Monthly Press, 2007), 157.

21. Smith, *Theory of Moral Sentiments*, 82.

22. Smith understood this, but he is often misinterpreted by later economists working from a more thoroughgoing, utilitarian, and individualistic mind-set. As James Halterman puts it: "Clearly Smith's notion of self-interest is not expressed as the isolated preference of an independent economic agent, but, rather, as the conditioned response of an interdependent participant in a social process." In "Is Adam Smith's Moral Philosophy an Adequate Foundation for the Market Economy?" *Journal of Markets and Morality* 6, no. 2 (Fall 2003): 459.

23. Robin Klay and John Lunn develop this idea in their excellent article "The Relationship of God's Providence to Market Economies and Economic Theory," *Journal of Markets and Morality* 6, no. 2 (Fall 2003): 547–59.

24. Adam Smith, *Theory of Moral Sentiments*, pt. 4, chap. 1.

25. Jim Wallis, *God's Politics* (San Francisco: HarperSanFrancisco, 2005), 263.

26. See, for instance, Paul Knitter and Chandra Musaffar, eds., *Subverting Greed: Religious Perspectives on the Global Economy* (Maryknoll, NY: Orbis, 2002).

27. *International Comparisons of Charitable Giving* (Kent, UK: Charities Aid Foundation, November 2006), www.cafonline.org/research.

28. As it happens, not only are Americans more generous than their industrial-world counterparts; a recent study by Arthur Brookes shows that in the United States, despite the rhetoric on the left, conservatives are more generous than liberals. This difference holds

when comparing the right and the left from the same income group. Arthur C. Brookes, *Who Really Cares: The Surprising Truth About Compassionate Conservatism* (New York: Basic Books, 2006).

29. George Gilder, *Wealth and Poverty* (San Francisco: ICS Press, 1993), 30.

30. Gilder, *Wealth and Poverty*, 20–24.

31. Gilder, *Wealth and Poverty*, 26.

32. For a compelling study of the importance of trust to economic and cultural success, see Francis Fukuyama, *Trust: The Social Virtues and the Creation of Prosperity* (New York: Free Press, 1996).

33. Gilder, *Wealth and Poverty*, 27.

34. Lowe's story is told in detail in John Duggleby, *The Entrepreneurial Journey of Edward Lowe* (Cassapolis, MI: Edward Lowe Foundation), 2006.

35. Gilder, *Wealth and Poverty*, 37.

36. Economist Israel Kirzner has emphasized the importance of discovery in the work of the entrepreneur. See his *Competition and Entrepreneurship* (Chicago: Univ. of Chicago Press, 1973).

37. Robert A. Sirico, *The Entrepreneurial Vocation* (Grand Rapids, MI: Acton Institute, 2001).

38. This is how George Gilder puts it in the documentary *The Call of the Entrepreneur* (Grand Rapids, MI: Acton Media, 2007).

39. Gilder, *Wealth and Poverty*, 33.

40. In the documentary *The Call of the Entrepreneur*.

CHAPTER 6: HASN'T CHRISTIANITY ALWAYS OPPOSED CAPITALISM?

1. Michael McTague, *The Businessman in Literature* (New York: Philosophical Library, 1979), 6–12.

2. John T. Noonan, *The Scholastic Analysis of Usury* (Cambridge, MA: Harvard Univ. Press, 1957), 20.

3. This doesn't mean that ancient peoples knew nothing of the market, or that basic economic realities of supply and demand didn't exist. But there are lower "transaction costs" when trading with family members than when trading with someone you don't know in another country. See Edd S. Noell, "A 'Marketless' World: An Examination of Wealth and Exchange in the Gospels and First Century Palestine," *Journal of Markets and Morality* 10, no. 1 (2007): 85–114.

4. There were rudimentary "banks" during the New Testament era and earlier, but they were much less significant economically than the banks we know of today.

5. Rodney Stark, *The Victory of Reason: How Christianity Led to Freedom, Capitalism, and Western Success* (New York: Random House, 2005), 112–13. I am indebted to Stark's lucid account in this section.

6. Stark, *Victory of Reason*, 113.

7. Stark, *Victory of Reason*, 45.

8. Samuel Gregg, *Banking, Justice, and the Common Good* (Grand Rapids, MI: Acton Institute, 2005), 47.

9. For a detailed study of the Reformers' views on usury, see David W. Jones, *Reforming the Morality of Usury: A Study of the Differences That Separated the Protestant Reformers* (Lanham, MD: Univ. Press of America, 2004).

10. Noonan, *Scholastic Analysis of Usury*, 399; Gregg, *Banking*, 38.

11. http://www.pinn.net/~sunshine/essays/wwjd.html.

12. Noonan, *Scholastic Analysis of Usury*, 2.

13. Max Weber, *The Protestant Ethic and the Spirit of Capitalism* (New York: Scribner's Sons, 1958). The book was originally published in 1904–1905. For a classic evaluation and critique of Weber's argument, see Christian socialist R. H. Tawney's book *Religion and the Rise of Capitalism: A History Study* (New York: Penguin, 1938).

14. Stark argues for the Christian origins of modern science in his previous book, *For the Glory of God: How Monotheism Led to Reformations, Science, Witch-Hunts, and the End of Slavery* (Princeton, NJ: Princeton Univ. Press, 2003).

15. For a detailed critique along these lines, see Samuel Gregg, "End of a Myth: Max Weber, Capitalism, and the Medieval World Order," *Journal des économistes et des études humaines* 13, nos. 2/3 (June/September 2003): 197–208. See also Samuel Gregg, *The Commercial Society* (Lanham, MD: Lexington Books, 2007), 3–21.

16. Stark, *Victory of Reason*, xiii.

17. Stark, *Victory of Reason*, 204–15.

18. George R. Beasley-Murray, *John*, Word Biblical Commentary, vol. 36 (Waco, TX: Word Books, 1987), 39.

19. Noell, "'Marketless' World," 102. Most biblical scholars think Jesus's act, falling when it does in his ministry, is meant to be a "prophecy of the temple's impending destruction." See Craig L.

Blomberg, *Neither Poverty nor Riches* (Downers Grove, IL: Inter-Varsity Press, 1999), 142–43.

20. This is clear from the wider context. Jesus introduces this and several other parables by saying: "Then the kingdom of heaven will be like this" (Matt. 25:1). He ends the parable with this mysterious warning: "For to all those who have, more will be given, and they will have an abundance; but from those who have nothing, even what they have will be taken away" (Matt. 25:29).

CHAPTER 7: DOESN'T CAPITALISM LEAD TO AN UGLY CONSUMERIST CULTURE?

1. Wallis, *God's Politics*, 355.
2. Rod Dreher, *Crunchy Cons* (New York: Crown, 2006), 29.
3. Dreher, *Crunchy Cons*.
4. Ron Sider frequently uses this distinction in *Rich Christians*.
5. For a careful and detailed discussion of these points, see John R. Schneider, *The Good of Affluence: Seeking God in a Culture of Wealth* (Grand Rapids, MI: Eerdmans, 2002).
6. Augustine of Hippo, *Late Have I Loved Thee*, ed. John F. Thornton (New York: Random House, 2006), bk. 2, pp. 5, 10.
7. U.S. Securities and Exchange Commission filings.
8. The only way to answer this question decisively would be to create some objective measure of gluttonous consumerism and compare the people in capitalist societies with those in noncapitalist societies. (For questions about consuming resources, see the next chapter.) And even if we could do that, the most we could detect would be a correlation between consumerism and capitalism. From the correlation alone we wouldn't know if one caused the other any more than knowing that the Hudson River rises with increased peanut consumption in the spring means one caused the other. To insist otherwise is known among logicians as the *post hoc* fallacy.
9. Dreher, *Crunchy Cons*, 22.
10. See Daniel Bell, *The Cultural Contradictions of Capitalism* (New York: Basic Books, 1976), 21–22. I think Bell links capitalism and modernism together too closely. The problem he identified is in modern materialism and the other -isms—nihilism, relativism, skepticism. None of these is the same as capitalism, which, as we have seen, buried its roots much earlier than the modern period.
11. For an excellent study of such worries among nineteenth- and twentieth-century intellectuals, see Jerry Z. Muller, *The Mind and*

the Market: Capitalism in Modern European Thought (New York: Knopf, 2002).

12. Dreher, Crunchy Cons, 221.

13. E. F. Schumacher, Small Is Beautiful (London: Hartley and Marks, 2000). The book was originally published in 1973.

14. Bill McGibben, Deep Economy: The Wealth of Communities and the Durable Future (New York: Times Books/Henry Holt, 2007). A popular recent argument for buying locally is that it's better for the environment, since it doesn't involve, say, shipping produce overseas. But it's not always so simple. See "Food Politics," Economist, December 7, 2006.

15. There's even a Web site, Delocator (http://www.delocator.net/index.php), where you can type in your zip code and find "independent," locally owned coffee shops, bookstores, and movie theaters. Why would you want to do that? The Web site explains: "Corporate industries invading American neighborhoods, from coffee chains to bookstore chains, music chains and movie theatre chains, pose a threat to the authenticity of our unique neighborhoods" (http://www.delocator.net/whydelocate.htm).

16. Martin Wolf offers a similar illustration in his Why Globalization Works (New Haven, CT: Yale Nota Bene, 2005), 3.

17. Wolf, Why Globalization Works, 44–45.

18. Paul Adams, "100-Mile Suit Wears Its Origins on Its Sleeve," Wired, March 30, 2007, http://www.wired.com/culture/design/news/2007/03/100milesuit0330.

19. See the picture at http://blog.wired.com/monkeybites/2007/03/photos_of_the_1_1.html.

20. Thomas Sowell, Basic Economics (New York: Basic Books, 2004), 85.

21. http://www.walmartfacts.com/content/default.aspx?id=3.

22. Quoted in Sowell, Basic Economics, 86.

23. http://www.pitachips.com/.

24. A few of the problems with ethanol: it costs more than oil to produce, it demands the use of fossil fuels to produce, it increases the price of corn and other agricultural products for rich and poor alike, and it leads to the increased use of herbicides and pesticides. Ethanol competes not through the free market but through special-interest lobbying and government protection.

25. For a critique of the 2007 farm bill, see Brian Riedl, "The Senate Farm Bill: A Missed Opportunity," Heritage Foundation Web-

Memo 1690 (November 5, 2007), http://www.heritage.org/Research/Agriculture/wm1690.cfm.

26. Rod Dreher can't seem to keep this matter straight in his *Crunchy Cons*.

27. Of course, some view technology itself as evil. This seemed to be the view of the Christian philosopher Jacques Ellul (1912–1994). In several books, Ellul developed the idea of the "technological tyranny over humanity." His 1967 book, *The Technological Society* (New York: Vintage, 1967), is sometimes considered a classic. He thought that technology was dehumanizing, in part because the artificiality "eliminates or subordinates the natural world." In my view, Ellul was unduly influenced by fashionable non-Christian philosophy such as neo-Marxism, and had a sub-Christian understanding of human creativity and dominion. Technology is an extension of human creativity. Human beings are part of the natural order, so there is nothing intrinsically alienating about technology.

28. This is how it reads on the back cover of *Crunchy Cons*. There is a different, more nuanced version inside the book. I discuss that later on.

29. A more scholarly book that discusses these issues in debate format is David Schindler and Doug Bandow, eds., *Wealth, Poverty, and Human Destiny* (Wilmington, DE: ISI Books, 2003).

30. Incidentally, "efficiency" is a shibboleth of critics of capitalism who haven't done their homework. To reduce the case for capitalism to efficiency is a grotesque caricature. In reality, very few defenders of the free market do so merely because the market is more efficient than the alternatives. Read F. A. Hayek, Ludwig von Mises, and Milton Friedman (to pick three free-market economists at random), and you'll see little about efficiency. You will read a great deal about the importance of liberty, justice, and limited government.

31. Joseph E. Schumpeter, *Capitalism, Socialism, and Democracy* (London: Routledge, 2006). The book was originally published in 1942.

32. For excellent how-to guides to enjoying small-scale, diversified, part-time farming, see any book on the subject written by Gene Logsdon. Logsdon is no fan of big agribusiness, but most of his arguments are refreshingly free-market friendly, urging Americans to stop subsidizing corn and other special-interest agriculture and start allowing smaller-scale pasture farmers to compete on a level playing field. For

a tonier, more self-righteous, antimarket approach to organic and "sustainable" agriculture, read up on the "Slow Food" movement at www.slowfood.com. It was founded by Italian Carlo Petrini in1989 and encapsulates many of the sentiments discussed in this chapter. It's official philosophical statement sets the tone:

We believe that everyone has a fundamental right to pleasure and consequently the responsibility to protect the heritage of food, tradition and culture that make this pleasure possible. Our movement is founded upon this concept of eco-gastronomy—a recognition of the strong connections between plate and planet.

Slow Food is good, clean and fair food. We believe that the food we eat should taste good; that it should be produced in a clean way that does not harm the environment, animal welfare or our health; and that food producers should receive fair compensation for their work.

We consider ourselves co-producers, not consumers, because by being informed about how our food is produced and actively supporting those who produce it, we become a part of and a partner in the production process.

The movement even has a manifesto, "The Slow Food Manifesto":

Our century, which began and has developed under the insignia of industrial civilization, first invented the machine and then took it as its life model.

We are enslaved by speed and have all succumbed to the same insidious virus: *Fast Life*, which disrupts our habits, pervades the privacy of our homes and forces us to eat Fast Foods.

To be worthy of the name, *Homo Sapiens* should rid himself of speed before it reduces him to a species in danger of extinction.

A firm defense of quiet material pleasure is the only way to oppose the universal folly of *Fast Life*.

May suitable doses of guaranteed sensual pleasure and slow, long-lasting enjoyment preserve us from the contagion of the multitude who mistake frenzy for efficiency.

Our defense should begin at the table with *Slow Food*.

Let us rediscover the flavors and savors of regional cooking and banish the degrading effects of *Fast Food*.

In the name of productivity, *Fast Life* has changed our way of being and threatens our environment and our landscapes. So *Slow Food* is now the only truly progressive answer.

That is what real culture is all about: developing taste rather than demeaning it. And what better way to set about this than an international exchange of experiences, knowledge, projects?

Slow Food guarantees a better future.

Slow Food is an idea that needs plenty of qualified supporters who can help turn this (slow) motion into an international movement, with the little snail as its symbol.

33. I can add to the list. I dislike the kitschy displays in shopping malls. I favor homeschooling and educational choice over the cookie-cutter public-school monopoly. I dislike new housing developments that seem to gratuitously flout the simplest principles of good design. I hate it when developers level mature trees and then replant when they're done.

34. Dreher, *Crunchy Cons*, 2.

35. There are several popular myths about "urban sprawl" and suburbs that I don't discuss here for lack of space. For discussion, see Ted Balaker and Sam Staley, "5 Myths About Suburbia and Our Car-Happy Culture," *Washington Post,* January 28, 2007, http://www.washingtonpost.com/wp-dyn/content/article/2007/01/26/AR2007012601589.html.

36. Ludwig von Mises, *The Anti-Capitalist Mentality* (Indianapolis, IN: Liberty Fund, 2006), 29.

37. Von Mises, *The Anti-Capitalist Mentality,* 29–30.

38. For a detailed discussion of these points, see Tyler Cohen, *In Praise of Commercial Culture* (Cambridge, MA: Harvard Univ. Press, 1998).

CHAPTER 8: ARE WE GOING TO USE UP ALL THE RESOURCES?

1. Stephen Moore, "Clear-Eyed Optimists," *Wall Street Journal,* October 5, 2007, w-11.

2. Paul Ehrlich, *The Population Bomb* (New York: Ballantine Books, 1968), xi.

3. Sider, *Rich Christians,* 157.

4. Sider, *Rich Christians*, 166.
5. Lynn Townsend White Jr., "The Historical Roots of Our Ecologic Crisis," *Science* 155, no. 3767 (March 10, 1967): 1203–7.
6. Thomas Sowell, *Basic Economics* (New York: Basic Books, 2004), 205.
7. Sowell, *Basic Economics*, 207.
8. For examples of historical prices of resources, see Julian L. Simon, *The Ultimate Resource 2* (Princeton, NJ: Princeton Univ. Press, 1996), 23–52. Simon discusses his famous bet with Ehrlich on pp. 32–33. See also Bjørn Lomborg, *The Skeptical Environmentalist: Measuring the Real State of the World* (Cambridge: Cambridge Univ. Press, 2001), 118–48.
9. Simon, *Ultimate Resource 2*, 164–67.
10. Sallie McFague, introduction to *Life Abundant: Rethinking Theology and Economy for a Planet in Peril* (Philadelphia: Augsburg/Fortress, 2000).
11. Edwin A. Abbott, *Flatland* (New York: Dover, 1952).
12. A classic nineteenth-century book, still in print, discussed this phenomenon: Charles Mackay, *Extraordinary Popular Delusions and the Madness of Crowds* (New York: Three Rivers Press, 1995).
13. One especially popular argument is that scientists and others who challenge the official story on global warming do so because of their financial interests. But this argument, in addition to being false, is a two-sided sword. If the paltry funding of skeptics taints their arguments, what of the billions of dollars of funding (and the sociological pressure it creates) for the other side? "Proponents of man-made global warming have been funded to the tune of $50 BILLION in the last decade or so, while skeptics have received a paltry $19 MILLION by comparison." "Newsweek's Climate Editorial Screed Violates Basic Standards of Journalism," http://epw.senate.gov/public/index.cfm?FuseAction=Minority.Blogs&ContentRecord_id=38d98c0a–802a–23ad–48ac-d9f7facb61a7.
14. Ellen Goodman, "No Change of Political Climate," *Boston Globe*, February 9, 2007. For more examples of such ad hominem arguments, see Jeff Jacoby, "Hot Tempers on Global Warming," *Boston Globe*, August 15, 2007, http://www.boston.com/news/globe/editorial_opinion/oped/articles/2007/08/15/hot_tempers_on_global_warming/.
15. Mike Allen, "Gore Wins Nobel Peace Prize," *Politico,* October 12, 2007, http://www.politico.com/news/stories/1007/6316.html.

16. *The Great Global Warming Swindle,* broadcast in the United Kingdom on Channel 4.

17. Quoted by Steven Milloy at http://www.junkscience.com/ByThe-Junkman/20070618.html.

18. James M. Taylor, "Warming Debate Not Over, Survey of Scientists Shows," Environment and Climate News (Heartland Institute: February 1, 2007), http://www.heartland.org/Article .cfm?artId=20512.

19. See Guillermo Gonzalez and Jay W. Richards, *The Privileged Planet* (Washington, DC: Regnery Publishing, 2004), 21–43.

20. For a tellingly biased report of this, see Kate Ravilious, "Mars Melt Hints at Solar, Not Human, Cause for Warming, Scientist Says," *National Geographic News,* February 28, 2007.

21. For a detailed study of global climate variations in geologic history and the possible causes, see Dennis Avery and Fred Singer, *Unstoppable Global Warming: Every 1,500 Years* (Lanham, MD: Rowman & Littlefield, 2007). ·

22. Bjørn Lomborg, *Cool It* (New York: Knopf, 2007).

23. Lomborg, *Cool It,* 40–43.

24. Jonah Goldberg, "Global Cooling Costs Too Much," *National Review,* February 9, 2007, http://article.nationalreview .com/?q=MmJiZDEyYzkxYWE0OWYxMWY4Y2ZjYzI2YmNmO GExMDE=.

25. The Copenhagen Consensus did a cost-benefit analysis to determine how best to spend $50 billion in humanitarian aid. Their top picks were projects to prevent HIV/AIDS, iron deficiency in women and children, and malaria. The Kyoto Protocol ranked sixteenth out of seventeen ways to spend the money. Moreover, they assumed that carbon dioxide is largely responsible for global warming. See the discussion at www.copenhagenconsensus.com. See also the compilation *Global Crises, Global Solution,* ed. Bjørn Lomborg (Cambridge: Cambridge Univ. Press, 2004).

26. See Lomborg, *Skeptical Environmentalist,* 3–33.

27. See Steven F. Hayward and Amy L. Kaleita, *Index of Leading Environmental Indicators,* 12th ed. (San Francisco: Pacific Research Institute; Washington, DC: American Enterprise Institute, 2007).

28. See Lomborg, *Skeptical Environmentalist,* 33.

29. For more statistics and charts than you can shake a stick at, see Simon, *Ultimate Resource 2,* 223–73. See also the worldwide demographic data at www.gapminder.com.

30. Jerome C. Glenn and Theodore J. Gordon, *2007 State of the Future* (New York: United Nations, 2007). The introductory sentence continues: ". . . but at the same time the world is more corrupt, congested, warmer, and increasingly dangerous. Although the digital divide is beginning to close, income gaps are still expanding around the world and unemployment continues to grow." See the executive summary at: http://www.millennium-project.org/millennium/sof2007-exec-summ.pdf.

31. Simon, *Ultimate Resource 2*, 233–73.

32. http://www.vhemt.org/.

33. Amateur scientist Forrest Mims broke the story in "Meet Doctor Doom," *Citizen Scientist*, March 31, 2006, http://www.sas.org/tcs/weeklyIssues_2006/2006–04–07/feature1p/index.html. Over the next few weeks, there was a media firestorm over the incident, and Mims was accused of misrepresenting Pianka's speech. I remain convinced that Mims basically got the story right.

The problem was that Pianka had asked that video cameras be turned off during his speech, and partial transcripts released later failed to fully corroborate Mims's account. But, as Mims pointed out, the transcript lacked precisely the part of the speech with the offensive comments. In any event, Mims's claim had several other corroborating pieces of evidence, which James Redford discusses in a blog posting titled "Forrest Mims Did Not Misrepresent Eric Pianka" (April 13, 2006, http://www.geocities.com/tetrahedronomega/pianka-mims.html). Cathy Young's piece in the *Boston Globe* focused the issue properly: the point was not that Pianka had called for the active extermination of 90 percent of the population. It's that he thought such extermination by natural causes (like the Ebola virus) would be a "good thing." Cathy Young, "Environmentalism and the Apocalypse," *Boston Globe*, April 17, 2006, http://www.boston.com/news/globe/editorial_opinion/oped/articles/2006/04/17/environmentalism_and_the_apocalypse/.

34. Natasha Courtenay-Smith and Morag Turner, "Meet the Women Who Won't Have Babies—Because They're Not Eco Friendly," *Daily Mail* (London), November 21, 2007, http://www.dailymail.co.uk/pages/live/femail/article.html?in_article_id=495495&in_page_id=1879.

35. "The Seventeen Percent Problem and Perils of Domestication," *New York Times*, August 13, 2007, http://www.nytimes.com/2007/08/13/opinion/13mon4.html?ei=5070&en=59c678f9636b6941&ex=1187668800&emc=eta1&pagewanted=print.

36. There are many complicated reasons for this. See Simon, *Ultimate Resource 2*, 311–56.
37. See United Nations Department of Economic and Social Affairs, *World Population to 2300* (New York: United Nations, 2004), http://www.un.org/esa/population/publications/longrange2/World-Pop2300final.pdf. Incidentally, there's no causal connection between population and poverty, except in the sense that you need people to have poor people. If there were no people, there would be no poverty. Hence the rigid logic of the population control crowd: get rid of most of the people.
38. Two recent documentary films illustrate this point: *The Call of the Entrepreneur* (Acton Media, 2007) and *The Ultimate Resource* (Free to Choose Media, 2007).
39. John Paul II, *Centesimus Annus* (1991), par. 32, http://www.vatican.va/holy_father/john_paul_ii/encyclicals/documents/hf_jp-ii_enc_01051991_centesimus-annus_en.html.

CONCLUSION: WORKING ALL THINGS TOGETHER FOR GOOD

1. Interestingly, the word for "providence" in Greek is *oikonomia*, from which we derive the word *economy*. *Oikonomia* initially referred to the management of a household. For this reason, Hayek disliked the use of the word *economy* to refer to the market.
2. Providence includes God's "omnipotence" and "omniscience." That is, God is all powerful and all knowing. More precisely, omniscience is something like God's power to know everything that it is logically possible for him to know. Omnipotence means that God can do anything that it is logically possible for him to do.
3. See this quote and further discussion in Robin Klay and John Lunn, "The Relationship of God's Providence to Market Economies and Economic Theory," *Journal of Markets and Morality* 6, no. 2 (2003): 541–64. This is one of the few published papers that consider the relationship of the market to the doctrine of providence.
4. Even a nonomniscient agent could plan an economy, if he had full and immediate access to all of those economic value judgments. Thus, an all-powerful and all-knowing God could providentially guide such a market order.
   Finite human beings, of course, are another matter. Since what's needed, in this case, is something like universal telepathic powers, we can be pretty sure that no human being will ever qualify for

the job. Therefore, for all practical purposes, Hayek's argument against socialist planning still stands.

5. For a detailed discussion of this argument, see the appendix.

APPENDIX: IS THE "SPONTANEOUS ORDER" OF THE MARKET EVIDENCE OF A UNIVERSE WITHOUT PURPOSE?

1. For a discussion of how Hayek's argument developed over time, see Bruce Caldwell, "Hayek and Socialism," *Journal of Economic Literature* 35, no. 4 (December 1997): 1856–90. For simplicity, I focus here on the most mature form of his argument, as found in Hayek, *Fatal Conceit*. This book was published just a few years before Hayek's death in 1992, and summarizes his argument against socialism.

2. He often refers to the market order as the "extended order" when speaking of the market. Of course, the idea of spontaneous order does not rest entirely with Smith and Hayek. It can be traced to Bernard Mandeville's *Fable of the Bees* (1712) and to Scottish Enlightenment figures such as David Hume and Adam Ferguson. The medieval Scholastics of Salamanca explored it even earlier. One might even trace the concept back to the Greek philosopher Heraclitus, who famously posited that order emerged from chaos and didn't need intelligence. Here, however, I am focusing only on Hayek's understanding and defense of the concept.

3. Hayek, *Fatal Conceit,* 24, 56, 67, 82, 107–8.

4. Hayek, *Fatal Conceit,* 8.

5. This phrase originated with Adam Ferguson (1767) in *An Essay on the History of Civil Science* (Edinburgh: Edinburgh Univ. Press, 1966): "Every step and every movement of the multitude, even in what are termed enlightened ages, are made with equal blindness to the future; and nations stumble upon establishments, which are indeed the result of human action, but not the execution of any human design" (pt. 3, sec. 2, p. 122).

6. Hayek does not seem to use his categories of "natural," "artificial," and "spontaneous" orders consistently. For discussion, see Elias L. Khalil, "Friedrich Hayek's Theory of Spontaneous Order: Two Problems," *Constitutional Political Economy* 8 (1997): 301–17.

7. A chaotic event corresponds to high rather than low entropy. And here things get a little confusing, because Claude Shannon, the founder of information theory, defined *information* in terms of entropy. In Shannon's sense, the less predictable and more surprising

a signal, the more information it contains. Like randomness, such "Shannon information" is complex rather than simple. Shannon's definition, however, does not distinguish mere randomness from meaningful patterns, such as written or transmitted messages. Much less does it distinguish meaningful patterns such as we find in messages from the order we find in physics.

8. Hayek, *Fatal Conceit*, 23–28.

9. Hayek's evolutionary account of our extended order is sophisticated, and I cannot do it justice here. But even Hayek admits that his account, like many just-so stories in evolutionary biology, is merely "conjectural history . . . of how the system might have come into being" (*Fatal Conceit*, 69). Such speculation can never be as compelling as the present existence of the market order itself, which is a manifest—observable—reality. Similarly, speculative history can never be as persuasive as Hayek's plausible argument against the human design of the market order. That argument takes the market order as a given, adds premises about the subjective nature of economic value and the limits of human knowledge, and concludes that no human could have designed it. There is nothing in this argument that conflicts with a purposeful view of the universe, much less rules out such a view. So where is the problem? It appears in Hayek's speculative account of how the market order emerged, where he introduces the dubious premise that order emerges from chaos.

10. Hayek, *Fatal Conceit*, 45.

11. Of course, some cosmologists find the theological implications of a low-entropy initial state unsettling. As a result, they are seeking a loophole that preserves materialism. But so far, no loophole has been found. For discussion, see Guillermo Gonzalez and Jay Richards, *The Privileged Planet: How Our Place in the Cosmos Is Designed for Discovery* (Washington, DC: Regnery, 2004), chaps. 9 and 10.

This may sound controversial, but even leading scientists like Francis Collins, head of the Human Genome Project, have suggested as much. See Francis Collins, "Faith and the Human Genome," *Perspectives on Science and Christian Faith 55*, no. 3 (September 2003): 152, http://www.asa3.org/ASA/PSCF/2003/PSCF9–03Collins.pdf.

One may think that phenomena such as hurricanes and tornadoes are examples of order emerging from chaos, since the weather patterns from which these orderly structures emerge seem chaotic. But this is merely a loose way of speaking. Tornadoes and hurricanes appear only under very specific initial conditions. Those conditions rest on so many variables that we cannot predict the emergence of these patterns with precision. But that gives us no reason to conclude that they literally emerge from chaos. Fractals are also sometimes cited as examples of complexity emerging from simplicity. But again, fractals are not order emerging from chaos. Fractals are patterns that reveal greater and greater complexity at higher and higher levels of resolution. But the patterns tend to be repeating, being generated (with computers) not from chaos, but from specific algorithms that run over and over using a feedback loop. Not only are fractals not an example of order emerging from chaos; they are also not a good analogy to the market order, since markets don't follow simple algorithms but instead emerge from the decisions of millions or even billions of agents.

12. Quoted in George Gilder, "Evolution and Me," *National Review*, July 17, 2006.
13. See, for instance, Richard Dawkins, *The Blind Watchmaker: Why the Evidence of Evolution Reveals a Universe Without Design*, reissue ed. (New York: W. W. Norton, 1996).
14. I can only assert rather than defend this idea here. But that is enough, since Hayek shared, in an inchoate form, this same antireductionist conviction. As a result, it is appropriate to criticize him when he violates his own principle. For a discussion of the hierarchical nature of the universe, see Nobel Prize–winning physicist Robert B. Laughlin's book *A Different Universe: Reinventing Physics from the Bottom Down* (New York: Basic Books, 2005) and Roger Penrose, *The Road to Reality: A Complete Guide to the Laws of the Universe* (New York: Knopf, 2005).
15. Hayek, *Fatal Conceit*, 127. Hayek was a staunch critic of scientism and reductionism, especially when the subject was human beings and their economic actions. As a result, he resisted the temptation to model economic reality after mathematical physics. He recognized that the market order is on the other end of the spectrum of the things we refer to as "orders."
16. Hayek, *Fatal Conceit*, 95.
17. Hayek, *Fatal Conceit*, 98. Hayek intuited that human agency could not be reduced to physics, but his default metaphysical resources could not accommodate that intuition.

18. Although I will not pursue the point here, I suspect Hayek's anti-reductionism would have been more at home in a theistic rather than a materialistic view of the world.

19. Hayek, *Fatal Conceit*, 83.

20. The principle of getting order from chaos is much like the principle of getting something from nothing. It's far too permissive. When either principle is allowed as a premise in an argument, virtually any conclusion can be derived. If Hayek's argument for a free market and against a planned economy depended on order emerging from chaos, that alone would be sufficient reason to reject his argument. Fortunately, his argument does not require that premise.

21. Hayek notes that the market order is "surprising." *Fatal Conceit*, 87.

22. Hayek frequently takes his critique of planning by human agents with their limited knowledge as a critique of such planning by any possible agent: "Thus socialist aims and programmes are factually impossible to achieve or execute: and they also happen, into the bargain as it were, to be logically impossible" (*Fatal Conceit*, 7). This is a mistake.